BOTANICAL RESEARCH AND PRACTICES

CHLOROPHYLL

STRUCTURE, PRODUCTION AND MEDICINAL USES

BOTANICAL RESEARCH AND PRACTICES

Additional books in this series can be found on Nova's website under the Series tab.

Additional E-books in this series can be found on Nova's website under the E-book tab.

MARINE BIOLOGY

Additional books in this series can be found on Nova's website under the Series tab.

Additional E-books in this series can be found on Nova's website under the E-book tab.

BOTANICAL RESEARCH AND PRACTICES

CHLOROPHYLL

STRUCTURE, PRODUCTION AND MEDICINAL USES

**HUA LE
AND
ELISA SALCEDO
EDITORS**

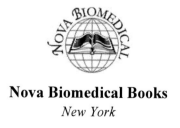

Nova Biomedical Books
New York

Copyright © 2012 by Nova Science Publishers, Inc.

All rights reserved. No part of this book may be reproduced, stored in a retrieval system or transmitted in any form or by any means: electronic, electrostatic, magnetic, tape, mechanical photocopying, recording or otherwise without the written permission of the Publisher.

For permission to use material from this book please contact us:
Telephone 631-231-7269; Fax 631-231-8175
Web Site: http://www.novapublishers.com

NOTICE TO THE READER

The Publisher has taken reasonable care in the preparation of this book, but makes no expressed or implied warranty of any kind and assumes no responsibility for any errors or omissions. No liability is assumed for incidental or consequential damages in connection with or arising out of information contained in this book. The Publisher shall not be liable for any special, consequential, or exemplary damages resulting, in whole or in part, from the readers' use of, or reliance upon, this material. Any parts of this book based on government reports are so indicated and copyright is claimed for those parts to the extent applicable to compilations of such works.

Independent verification should be sought for any data, advice or recommendations contained in this book. In addition, no responsibility is assumed by the publisher for any injury and/or damage to persons or property arising from any methods, products, instructions, ideas or otherwise contained in this publication.

This publication is designed to provide accurate and authoritative information with regard to the subject matter covered herein. It is sold with the clear understanding that the Publisher is not engaged in rendering legal or any other professional services. If legal or any other expert assistance is required, the services of a competent person should be sought. FROM A DECLARATION OF PARTICIPANTS JOINTLY ADOPTED BY A COMMITTEE OF THE AMERICAN BAR ASSOCIATION AND A COMMITTEE OF PUBLISHERS.

Additional color graphics may be available in the e-book version of this book.

Library of Congress Cataloging-in-Publication Data

Chlorophyll : structure, production and medicinal uses / editors: Hua Le and Elisa Salcedo.
    p. cm.
Includes index.
ISBN 978-1-61470-974-9 (hardcover)
1. Chlorophyll. I. Le, Hua. II. Salcedo, Elisa.
QK898.C5C45 2011
572'.59--dc23
                      2011028043

Published by Nova Science Publishers, Inc. † New York

Contents

Preface vii

Chapter I Impact of Molecular Organization on UV-Irradiation Effects to Chlorophyll Stability: A Base to Understand Biomedical Applications 1
Dejan Markovic and Jelena Zvezdanovic

Chapter II From Protochlorophillide to Chlorophyll: Final Light-Dependent Stage in Biosynthesis of Main Photosynthetic Pigment 43
Olga B. Belyaeva and Felix F. Litvin

Chapter III Genomics and Phenomics of Chlorophyll Associated Traits in Abiotic Stress Tolerance Breeding 81
U. R. Rosyara, N. K. Gupta, S. Gupta, and R. C. Sharma

Chapter IV Chlorophyll Fluorescence Emission Spectra in Photosynthetic Organisms 115
María Gabriela Lagorio

Chapter V Sedimentary Chlorophyll and Pheopigments for Monitoring of Reservoir Characterized by Exclusively High Dynamism of Abiotic Conditions 151
L. E. Sigareva and N. A. Timofeeva

Chapter VI	Medicinal Uses of Chlorophyll: A Critical Overview *V. K. Mishra, R. K. Bacheti and Azamal Husen*	**177**
Index		**197**

Preface

Chlorophyll is a green pigment found in almost all plants, algae and cyanobacteria, and is an extremely important biomolecule, critical in photosynthesis, which allows plants to obtain energy from light. In this book, the authors present current research in the study of the structure, production and medicinal uses of chlorophyll. Topics discussed include the impact of molecular organization on UV-irradiation effects to chlorophyll stability and biomedical applications; chlorophyll biosynthesis; genomics and phenomics of chlorophyll associated traits in abiotic stress tolerance breeding; chlorophyll fluorescence emission spectra in photosynthetic organisms and traditional medicinal uses of chlorophyll.

Chapter 1 - Use of chlorophyll (Chl) and Chl derivatives as the photosensitizers in few anti-cancer types of therapies, as well as a probes to monitor cancer-targeted type of cells and tissues has been continuing to attract research interest both on theoretical as well as on the practical side. Whatever of the method has been employed for this purpose they are all based on interaction of chlorophyll with the appropriate light, which spectrum profile and intensity characteristics depend on the particular method and it's postulated goals. In such a circumstances stability of chlorophyll molecule to the action of the applied light emerges as an important factor *per se*. Question of chlorophyll stability particularly starts to play an important and intriguing role under circumstances when the amount of incident excitation energy widely overwhelm a capacity of chlorophyll molecule to safely hand the excess of excitons, either through *intra-* and *inter-* radiative and non-radiative dissipative mechanisms, or through a productive and controlled photochemical conversion of its longer-lived triplet state. A special case of chlorophyll and it's derivatives use for therapeutic purposes are pharmaceutical and other biomedical formulations for various dermatological applications used against damaging action of a natural sunlight, which is generally and conventionally known as "skin aging". There, chlorophyll may be a part of PDT-drug, playing curative role, acting as a sensitizer which usually needs a chlorophyll "support counterpart", *i.e.* a chlorophyll-binding ligand molecule to facilitate Chl suitable incorporation inside the curing tissue. However Chl is also used as the ingredient of dermatological applications based on water-oil emulsions systems, where it acts as a skin restorer and refresher, healing (natural sunlight induced) lesions and burns (so, soothing consequences), while in the same time it may act preventively by absorbing part of sunlight visible spectrum that leads to appearance of skin rashes and burns (so, acting against the cause); in the latter (non-PDT case) it doesn't need the ligand(s). Chlorophyll absorb most efficiently in the visible (VIS) range of the natural sunlight spectrum (400-800 nm) and less efficiently in ultraviolet (UV) range. Natural sunlight always contains a certain portion of UV-A fraction (320-400 nm), but also a small percentage of more energetic UV-B fraction (290-320 nm) which has been dramatically increased in the last two decades due to a widespreading of the ozone hole. Though it's contribution to the overall natural sunlight content does not exceed few percents UV-B produces the biggest portion of the serious damage, by initializing mechanism of free radicals, especially oxygenic (ROS) radicals production that consequently leads to a lot of pathological consequences on a short, as well as on a longer time scale. One of the usual ways to at least partly prevent this damage – done especially by mutual synergetic and synchronous UV-A and UV-B action inside the whole sunlight spectrum – is to employ numerous efficient UV-absorbers in so-called (dermatological) sun-screen preparations. However, it does not prevent appearance of lesions and burns induced by VIS-part of the sunlight spectrum, which rises a requirement of an absorber which (absorption) spectrum matches the whole UV (UV-A and UV-B)-VIS range of the sunlight emission spectrum, and chlorophyll appears as a suitable choice to meet this request. In this view chlorophyll response to UV-part of the sunlight spectrum (and to UV overall) emerges as a problem *per se*, and needs a particular elaboration. Of course chlorophyll response to increasing intensities of sunlight (particularly in the summer period), and especially to a UV-A and UV-B portion of it is not a simple and straightforward. The interaction depends on many factors, structural composition and molecular organization of the surrounding environment (water-oil based emulsion) being one of the crucial ones. The lipoidal part of the formulations is proned to (UV-induced) oxidation and peroxides formation *via* either free radicals chain mechanism or the non-radical pathway where chlorophyll may play a sensitizer role. The newly created radicals, and generally the ROS species, in turn may additionally affect chlorophyll stability to UV-action. So, to get deeper insight into a chlorophyll response as an ingredient of the dermatological formulations one should look at the response data in a very simple systems (solutions), in a model systems with increasing levels of molecular organizations (biological models) and in the isolated organelles where chlorophylls act in the natural environment. This contribution pursues this approach.

Chapter II - The key step in chlorophyll biosynthesis is photoreduction of its immediate precursor, protochlorophyllide. This reaction consists in the attachment of two hydrogen atoms in positions C17 and C18 of the tetrapyrrole molecule of protochlorophyllide; the double bond is replaced with the single bond. The purpose of this survey is to summarize and discuss the data obtained in the studies on organization of the active pigment–protein complex where the photoreduction of protochlorophyllide proceeds, the mechanisms of the primary photophysical and photochemical reactions of this process and the pathways of functionally chlorophyll species formation. The active pigment–protein complex includes protochlorophyllide, the donor of hydride ion NADPH, and the photoenzyme protochlorophyllide oxidoreductase (POR). In the living cell there are several spectrally different active forms of the chlorophyllide precursor. Based on the results of numerous investigations, it can be stated that the reduction of the active forms of protochlorophyllide is a multistep process comprising two or three short-living intermediates characterized by the singlet ESR signal. The first intermediate seems to be a complex with a charge transfer between protochlorophyllide and the donor of hydride ion (NADPH). The donor of the second proton is the tyrosine residue Tyr 193 of the photoenzyme. The photoreduction of protochlorophyllide is preceded by light-stimulated conformational changes in the enzyme active site, enabling the hydride and proton transfer reactions to occur. Formed as a result of the photoreaction, the primary forms of chlorophyllide undergo further dark and light-dependent transformations, sequential and parallel, leading to the formation of different forms of chlorophyll and pheophytin that start the formation of the pigment apparatus of the two photosystems and light-harvesting antenna.

Chapter III - Development of stress tolerance in economically important crops is very important in context of recent challenges brought by changing climate along with increased demand for both food and fuel. Crop plants are affected by several abiotic factors such as high or low temperature, excessive water or drought conditions, low or high light, nutrient deficiency or nutrient excess, salt, pollutant and heavy metal. Studies have shown genetic variation in tolerance to these stresses and chlorophyll associated traits such as stay green, chlorophyll content, chlorophyll fluorescence and spectral reflectance are considered as indicator to stress tolerance to one or more of these stresses. Advancement in science of abiotic stress tolerance breeding relies on advancement in genomics and phenomics. The phenomics of stress tolerance are now focused on development of precision high throughput techniques whereas genomics has advanced the authors' knowledge on genes or quantitative trait loci (QTL) allowing selection to perform at gene level and production of genetically engineered crops. This review article presents an insight into practical application of physiology and molecular biology of chlorophyll alternation in response to stresses in context of development of novel stress tolerant plant genotypes.

Chapter IV - The description, analysis and applications of chlorophyll fluorescence emission spectrum in biological organisms are reviewed. At room temperature, in photosynthetic tissues, chlorophyll fluorescence presents a peak in the red (about 680 nm) due to emission from chlorophyll-a linked to photosystem II and in the far-red (about 735 nm) due to contribution of both photosystems (I and II). Chlorophyll fluorescence analysis is a powerful tool for plant physiologists but its interpretation is usually complicated by the presence of light re-absorption processes inside the photosynthetic tissue. A great deal of important conclusions on photochemistry of plants is usually inferred from leaves fluorescence ratios at different emission wavelengths (typically the fluorescence ratio red/far-red). Nevertheless, most of them are deduced from observed spectra distorted by light re-absorption processes and the resultant conclusions are not reliable. A review of the empirical and theoretical approaches to evaluate the chlorophyll emission spectra inside the photosynthetic organism, free from light re-absorption processes, published to date in literature, is detailed and discussed in this chapter. Not only the steady-state spectrum from plant leaves but also from algae and chlorophyll-containing fruits are discussed. Applications in plant physiology and in monitoring plant stress are presented.

Chapter V - Research was made to substantiate the use of sedimentary pigments for monitoring of a trophic state of a large reservoir characterized extremely mosaic structure of bottom sediments and rare stratification. Spectrophotometric method was used to measure the concentrations of sedimentary pigments (chlorophyll *a*, pheopigments, total carotenoids) in the surface sediments (0–2.5 and 2.5–5.0 cm) of the Rybinsk reservoir, Russia. Bottom sediments were sampled at 6 permanent stations and from 22 to 43 stations of episodic observations at river and lake-like sites of the Rybinsk reservoir in 1993–2010. Indexes E_{480}/E_{665} and $E_{480}/(1.7E_{665acid})$ and per cent pheopigments were considered as indicators of degradation of pigment fund. Spatial (horizontal) and temporal (seasonal and long-term) distributions of sedimentary pigments in relation with water depth and temperature, Secchi depth, sediment types and concentrations of phytoplankton chlorophyll *a* were studied. The concentrations of sedimentary chlorophyll *a* and pheopigments (Chl+Pheo) most frequently found in upper 2.5 cm sediment layer for 1993–2010 were in the range of mesortrophic and eutrophic values. The mean concentrations of Chl+Pheo for ice-free period at the separate stations varied in the range 3–284 µg/g dry matter Average of mean annual Chl+Pheo concentrations in 1993–2010 was maximum (167.4 µg/g dry matter) at ecoton site with sandy silts and clay silts, and minimum (28.4 µg/g dry matter) in lake-like part with mosaic sediments where sand was dominated. Despite strong heterogeneity of water masses and sediment complex, the positive dependence between chlorophyll concentrations in water column and concentrations of Chl+Pheo in surface sediments of the reservoir was established. This dependence reflects the phytoplankton role in formation of productivity of bottom biotopes. Mean values of the ratio of chlorophyll content in water column to concentrations of Chl+Pheo in bottom sediments were compared with sediment accumulation rates, calculated using data of bottom sediment investigations. The mean concentration of sedimentary Chl+Pheo in the reservoir (upper 2.5 cm layer) was calculated for two periods taking account of areas of different type sediments. For 1996–1998 it amounted to 37.0±8.5 µg/g dry matter or 15.3±2.4 mg/(m²·mm fresh matter) and for 2009–2010 – 28.1±7.5 µg/g dry matter or 10.4±3.8 mg/(m²·mm fresh matter). Decrease in sedimentary pigment content in recent years is in agreement with conception of reservoir de-eutrophication on the basis of other hydrological and hydrobiological data. In a year with extremely hot summer and long calm weather (2010) the mean concentrations of Chl+Pheo at the stations did not differ statistically from those in normal (2009) year. However, in 2010 indexes E_{480}/E_{665} and $E_{480}/(1.7E_{665acid})$ decreased, i.e. degree of degradation of sedimentary pigment fund decreased. It was assumed that in 2010 preconditions for increase in phytoplankton productivity in the future were created.

Chapter VI - Reports on traditional medicinal uses of chlorophyll in alternative forms of medicine are known since ages. Now-a-days chlorophyll has been used in the field of medicine as remedy and diagnostics. Chlorophyll molecules are used in pharmacy as photosensitizer for cancer therapy. Their roles as modifier of genotoxic effects are becoming increasingly important, besides these it being known to have multiple medicinal uses. Chlorophyll has its place in modern medicine. Here, the authors present a review of recent developments in medicinal uses of chlorophyll. This article enumerates therapeutic claims of chlorophyll as drugs based on investigative findings of modern science. A brief overview of research and developments of medicinal uses of chlorophyll will be presented in this review along with challenges of potential applications of chlorophyll and its derivatives as chemotherapeutic agents.

In: Chlorophyll
Editors: H. Le, et.al.

ISBN: 978-1-61470-974-9
© 2012 Nova Science Publishers, Inc.

Chapter I

Impact of Molecular Organization on UV-Irradiation Effects to Chlorophyll Stability: A Base to Understand Biomedical Applications

Dejan Markovic and Jelena Zvezdanovic
Faculty of Technology, University of Nish, Leskovac, Serbia

Abstract

Use of chlorophyll (Chl) and Chl derivatives as the photosensitizers in few anti-cancer types of therapies, as well as a probes to monitor cancer-targeted type of cells and tissues has been continuing to attract research interest both on theoretical as well as on the practical side. Whatever of the method has been employed for this purpose they are all based on interaction of chlorophyll with the appropriate light, which spectrum profile and intensity characteristics depend on the particular method and it's postulated goals. In such a circumstances stability of chlorophyll molecule to the action of the applied light emerges as an important factor *per se*. Question of chlorophyll stability particularly starts to play an important and intriguing role under circumstances when the amount of incident excitation energy widely overwhelm a capacity of chlorophyll molecule to safely hand the excess of excitons, either through *intra-* and *inter-* radiative and non-radiative dissipative mechanisms, or through a productive and controlled photochemical conversion of its longer-lived triplet state.

A special case of chlorophyll and it's derivatives use for therapeutic purposes are pharmaceutical and other biomedical formulations for various dermatological applications used against damaging action of a natural sunlight, which is generally and conventionally known as "skin aging". There, chlorophyll may be a part of PDT-drug, playing curative role, acting as a sensitizer which usually needs a chlorophyll "support counterpart", *i.e.* a chlorophyll-binding ligand molecule to facilitate Chl suitable incorporation inside the curing tissue. However Chl is also used as the ingredient of dermatological applications based on water-oil emulsions systems, where it acts as a skin restorer and refresher, healing (natural sunlight induced) lesions and burns (so, soothing consequences), while in the same time it may act preventively by absorbing part of sunlight visible spectrum that leads to appearance of skin rashes and burns (so, acting against the cause); in the latter (non-PDT case) it doesn't need the ligand(s). Chlorophyll absorb most efficiently in the visible (VIS) range of the natural sunlight spectrum (400-800 nm) and less efficiently in ultraviolet (UV) range. Natural sunlight always contains a certain portion of UV-A fraction (320-400 nm), but also a small percentage of more energetic UV-B fraction (290-320 nm) which has been dramatically increased in the last two decades due to a widespreading of the ozone hole. Though it's contribution to the overall natural sunlight content does not exceed few percents UV-B produces the biggest portion of the serious damage, by initializing mechanism of free radicals, especially oxygenic (ROS) radicals production that consequently leads to a lot of pathological consequences on a short, as well as on a longer time scale. One of the usual ways to at least partly prevent this damage – done especially by mutual synergetic and synchronous UV-A and UV-B action inside the whole sunlight spectrum – is to employ numerous efficient UV-absorbers in so-called (dermatological) sun-screen preparations. However, it does not prevent appearance of lesions and burns induced by VIS-part of the sunlight spectrum, which rises a requirement of an absorber which (absorption) spectrum matches the whole UV (UV-A and UV-B)-VIS range of the sunlight emission spectrum, and chlorophyll appears as a suitable choice to meet this request. In this view chlorophyll response to UV-part of the sunlight spectrum (and to UV overall) emerges as a problem *per se*, and needs a particular elaboration.

Of course chlorophyll response to increasing intensities of sunlight (particularly in the summer period), and especially to a UV-A and UV-B portion of it is not a simple and straightforward. The interaction depends on many factors, structural composition and molecular organization of the surrounding environment (water-oil based emulsion) being one of the crucial ones. The lipoidal part of the formulations is proned to (UV-induced) oxidation and peroxides formation *via* either free radicals chain mechanism or the non-radical pathway where chlorophyll may play a sensitizer role. The newly created radicals, and generally the ROS species, in turn may additionally affect chlorophyll stability to UV-action. So, to get deeper insight into a chlorophyll response as an ingredient of the dermatological formulations one should look at the response data in a very simple systems (solutions), in a model systems with increasing levels of molecular organizations (biological models) and in the isolated organelles where chlorophylls act in the natural environment. This contribution pursues this approach.

1. Introduction

1.1. Chlorophyll – The Basics

Chlorophyll (Chl) is a chlorin, porphyrin derivative, a cyclic tetrapyrrole with isocyclic cyclopentanone ring, fused at the edge of the right-bottom pyrrole ring; the central Mg-atom plays a coordinating role (Scheer 1991) (Figure 1). Major functions of Chl in photosynthesis are light-harvesting and light conversion in Photosystem I and II (PSI and PSII) inside photosynthetic apparatus (Scheer 1994; Scheer 2003a). The chlorophylls comprise a group of more than 50 tetrapyrrolic pigments with common structural elements and function (Scheer 2003b; 2006). The most abundant Chl form, especially in plants and green algae is Chl$a$, usually followed by Chl$b$. Chlorophyll a itself may appear in various spectral forms, depending on the function it performs; for example, Chla_{680} is reaction trap of PSII reaction center, while Chla_{700} is reaction trap of PSI reaction center in green plants; the index numbers show the absorbing wavelengths (Scheer 2006).

It is necessary to add that chlorophyll – besides Chl$a$ and Chl$b$ – exist also in some very rare forms, like Chl$c$, - d, -e, -f, and the same is valid for chlorophyll of photosynthetic bacteria, Bacteriochlorophyll – BChl (Scheer 1991; 2006). Furthermore, Chl is also accompanied by presence of it's various derivatives in various photosynthesis "subjects" (plants, algae, bacteria…), like pheophytins ("demetalated chlorophylls", or simply, chlorophylls without central Mg-atom), chlorophyllides (Chl without phytol tail), pheophorbides (demetalated chlorophyllides) – and their derivatives – as the most important among them (Hynninen 1991; Scheer 2006).

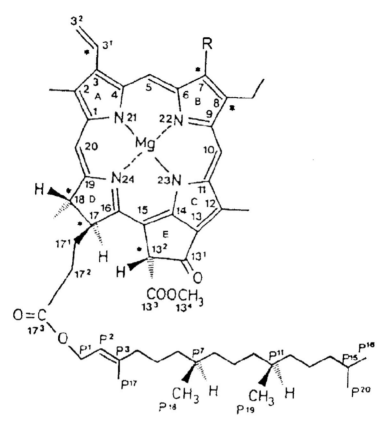

Figure 1. Chlorophyll structure, constituting from porphyrin ring and phytol hydrocarbon chain. The C-atoms are numerated according to IUPAC nomenclature rules. R=CH3 in the case of Chl*a*, and CHO in the case of Chl*b*. The assymetrically substituited C-atoms are marked by star symbol (*).

As Figure 1 shows the central Mg-atom of Chl molecule is bonded to N-atoms of the four pyrroles rings (A-D) by two covalent and two coordinative bonds. The 5-membered isocyclic ring (E) is connected to porphyrin in C(13-15) positions, while the phytol-esterified residue of propionic acid is attached to the chlorin C-17^3 position.

It is worth to note that Chl may be considered as just one compound in an unlimited row of similar compounds that perform other very important biological functions (for example, heme components of some crucial proteins such are myoglobin or hemoglobin have very similar porphyrin structure to Chl). However, untill today, any „Chl-surogat" capable of performing major photosynthesis function has not been synthesized yet, despite advanced work on photosynthesis biological (artificial) models with more or less reconstructed structure of photosynthesis apparatus.

As chlorophylls are porphyrin derivatives one should expect similarities in their optical spectra (Hoff and Amesz 1991). One simple model which treats them is Gouterman model (1978), explaining porphyrin spectra as a result of linear combination of electron transitions between two highest populated and two lowest unpopulated molecular π-orbitals (Figure 2-A): the former ones (HOMO and HOMO-1) are a_{1u} and a_{2u} (π) orbitals, while the latter ones (LUMO and LUMO+1) belong to e_{1u} and e_{2u} (π) orbitals[1]. The substracting (minus) combination of the two orbitals yields Q-bands in red range of visible (VIS) spectra („red band"), while the plus (+) combination yields B- or Soret-band, or „blue band" (Figure 2-B). The two bands position is determined by the extended π-delocalization at the edge of cyclic tetrapyrrole (porphyrin) skeleton, creating "conjugated" type of structure and providing a huge batochromic shift in VIS range[2]. The Chl "red" band is assigned as Q_y-band, and the absorption maxima (A_{max}) for Chl$a$ and Chl$b$ in acetone are located at 662.1 nm and 645.5 nm, respectively (Jeffrey et al. 1996)[3]. Of course the bands intensities and the maximum absorption positions (λ_{max}) in the other solvents are different (Jeffrey et al. 1996), and they are also influenced by many other factors like substitution, ligands, H-bonding, the surrounding conditions (in vitro, in situ, in vivo). The ratios of absorbance intensities for Soret- and the Q_y-band are 1.3 for Chl$a$, and 2.8 for Chl$b$ (Kobayashi et al. 2006).

When chlorophyll is attached to the appropriate ligand the λ_{max} position may suffer an additional "red" shift. A special case of this type is attributed to Chl-organization in isolated photosynthetic organelles (chloroplasts) and sub-organelles (thylakoids) where Chls are aggregated and placed inside lipo-protein matrix[4]. This leads to additional extension of π-conjugated system, moving further "red" both Soret and Q_y-band; the λ_{max} for Chl$a$ is now at 680 nm, while for Chl$b$ at 650 nm (Scheer 2006).

---

[1] The assignements for the electronic states symmetry – *a* i *e* – are related to the whole molecule symmetry. The porphyrin type of molecules with D_{4h} symmetry have molecular orbitals which eletronic states are labelled as a_{1u}, a_{2u}, e_{gx} i e_{gy}.

[2] The "conjugated" type of structure means alternated succession of single and double bonds.

[3] A very week band in a form of shoulder at about 615 nm probably comes from higher overtones combinations of Q_y-transition, while a week band with a maximum at about 578 nm is assigned as Q_x-band.

[4] Here, the abbreviation "Chls" instead "Chl" has been used, and that will be repeated occasionally through the whole text. "Chls" means "all possible chlorophylls" in the particular case, so not only Chl$a$, or just Chl$a$ and Chl$b$, but also other Chl forms (like Chl$c$) and Chl-derivatives (pheophytins, chlorophyllides, pheophorbides), too; when only "Chl" is used, it means either Chl$a$ (sometimes Chl$a$ and Chl$b$), or just Chl-type of structure.

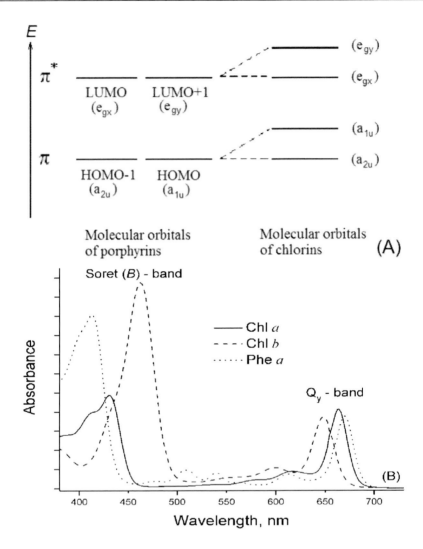

Figure 2. (A) Energy levels of π-molecular orbitals of porphyrins (left), and chlorins (right – including chlorophylls); the electronic transitions take place between them, following photons absorption in visual (VIS) range. π – bonding molecular orbitals of the highest populated energetic levels of porphyrin and chlorin rings (HOMO); π* - antibonding molecular orbitals of the lowest unpopulated energetic levels of porphyrin and chlorin rings (LUMO) (B) Absorption spectra of major photosynthetic pigments (the chromophores of the 3rd generation photosensitizers): chlorophyll a (———), chlorophyll b (- - - - -), pheophytin a (·······). The "diagnostic" or "blue photon band" (also, Soret, or B-band), originating from S_2 excited states, are on the left side, whereas the "curative" or "red photon band" (Q_y), originating from S_1 excited states, are on the longer wavelengths side of the spectra.

1.2. Chlorophyll Photosensitization as a Base for It's Use in Medicine

Chlorophyll relevance for health, in a very broad sense, was known long time ago. Even today health-related studies dealing with chlorophyll and done on tested animals do not unconditionally require modern technologies to support this knowledge. For example as a part of specific diet chlorophyll may inhibit tumorogenesis in rainbow trout (Simonich *et al.* 2008) as it inhibits aflatoxin B1-induced multiorgan carcinogenesis in rats (Simonich *et al.* 2007). A recently published experiment done on human (female) population prove that Chl intake in the form of Chl extract improves signs of skin photoaging and increases levels of Type I Procollagen in the skin (Cho *et al.* 2006).

Today most advanced use of chlorophyll in medicine, *based on Chl interaction with light*, is certainly related with PhotoDynamic Therapy – PDT (biomedical methods that use other characteristics of chlorophyll – other than light sensitivity – will not be discussed in this review).

1.2.1. Chlorophyll, Photosensitizers and PDT

Photodynamic therapy of tumors has become one of the leading methods of selective killing of cancer cells. For the last two decades it has passed a long way from basic photochemical and photophysical studies in a simple, homogeneous medium (solutions etc.) to a clinical treatment of patients suffering of various types of cancers. PDT is based on a photochemical process for producing localized tissue necrosis, which involves the activation of a photosensitizing drug in the target tissue with light of a specific wavelength matched to an absorption peak of the photosensitizer in the presence of molecular oxygen (Dougherty *et al.* 1998; Berg *et al.* 2005).

Central role in PDT belongs to photosensitizers. A photosensitizer is generally defined as a substance inducing light sensitivity to a normally light insensitive chemical or physical process. For a successful PDT therapy it is important to know the photophysical properties of the photosensitizer molecules. The studies of these properties have to be carried out at different levels, in a medium of different degrees of molecular organization. The first step is determination of the photophysical properties of "monomeric" dye molecules (Roeder 1998). The next step is a study on the photosensitizer behavior in microheterogeneous, *i.e.* water-containing environments. The goal of these investigations is to clarify whether the dyes molecules undergo to aggregation in such environments or not (*e.g.* whether are converted into "oligomer" forms). The systems studied usually include membrane modeling systems (micelles, liposomes, vesicles, erythrocyte ghosts) or carrier systems (liposomes, antibodies).

1.2.2. Photophysical Characterization of Sensitizer Molecules

Absorbing a photon, a dye molecule undergoes to excitation (femtosecond scale) to the first or higher excited singlet states (Jablonsky diagram – Figure 3-A). From higher excited states the exciton dissipates very fast (in picosecond range) to the first excited singlet state, having a lifetime of a few nanoseconds. From S_1 level the exciton is dissipated by a few competing mechanisms. The $S_1 \rightarrow S_0$ (the ground state) deactivation may occur in an irradiated form (fluorescence) or non-irradiated form (internal conversion). Alternatively, the exciton may undergo to intersystem crossing (*isc*) to the nearest triplet excited state, T_1. The T_1 state has a longer lifetime compared to it's singlet counterpart, and therefore is of primary photochemical interest, *i.e.* it is photochemically more reactive than any other excited state. However, the $S_1 \rightarrow T_1$ transition is forbidden by all-known selection rules (particularly, the multiplicity rule), and therefore the T_1 state is usually non-populated.

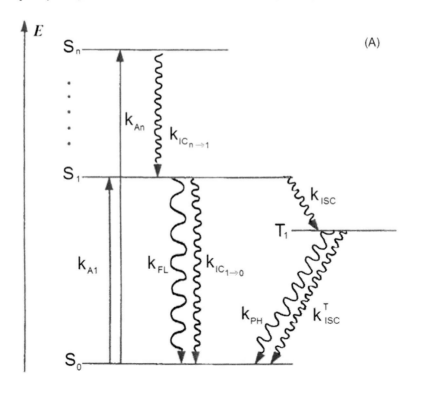

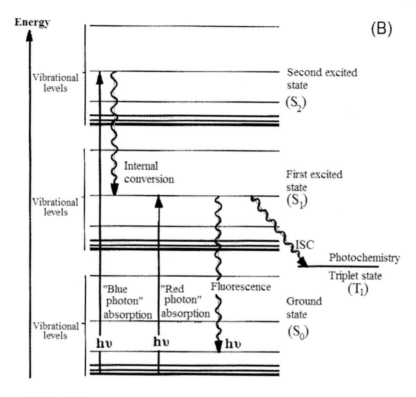

Figure 3. (A). Jablonsky-diagram of an excited dye molecule. S_0 - ground state, S_1 - first excited state; S_n - higher excited states; T_1 - first excited triplet state. k - rate constant. The subscripts meaning: A - absorption; IC - internal conversion; ISC - intersystem crossing; FL - fluorescence; PH - phosphorescence. Internal conversion and intersystem crossing are non-radiative dissipative exciton pathways, while fluorescence and phosphorescence represent radiative deactivation of S_1 and T_1 states, respectively.(B) Jablonsky-diagram for chlorophyll (Chl) molecule, *i.e.* mechanisms of energy dissipation of an excited Chl molecule. The absorption transitions (↑), $S_0 \rightarrow S_1$, $S_0 \rightarrow S_2$, have been created by absorption of „red" and „blue" photons, respectively. Non-radiative dissipation: internal conversion (*ic*) $S_2 \rightarrow S_1$, and intersystem crossing (*isc*), $S_1 \rightarrow T_1$. Radiative dissipation: fluorescence, $S_1 \rightarrow S_0$.

Fortunately for PDT, the macrocyclic molecules – like porphyrins, and so chlorophylls - are an exception because of the spin-orbital coupling. In such a case, the k_{isc} rate constants are not negligible, and the T_1 state is populated in μs-ms range. It is usually detected by a week phosphorescence emission (Wilson and Cerniglia 1994) (Figure 3-A), but more important, it is a starting point for a photosensitation reaction.

The photosensitation reaction usually occurs by two main competing pathways (Foote 1976, 1991; Girotti 2001), but in both cases it implies the presence of a neighboring molecule reacting with the photosensitizer's long-lived triplet state. The first pathway (the Type I mechanism) is a free-radical mechanism which may be accompanied by proton transfer (*e.g.* hydrogen abstraction) or electron transfer. The other one is non-radical mechanism (Type II), implying the energy transfer from T_1 photosensitizer state to oxygen in it's ground triplet state, $^3O_2$, yielding highly reactive singlet oxygen, $^1O_2$:

$$^3Sens + {}^3O_2 \rightarrow {}^1Sens + {}^1O_2 \tag{1}$$

Both processes are running in competition. The efficiency depends on the surroundings and the nature of the substrate molecules.

In the presence of molecular oxygen the Type II mechanism becomes more dominant. The very well known high toxicity of the singlet oxygen has been employed in photodynamic therapy for killing of cancer cells. Since the activation energy of $^1O_2$ is 0.96 eV, that means that the $S_0 \rightarrow S_1$ absorption of the photosensitizer must be at wavelengths of no longer than 850 nm (Roeder 1998). Sensitizers absorbing at longer wavelength show strongly reduced $^1O_2$ yields due to reversibility of energy transfer (Moser 1998).

There are a lot of photosensitizers used today in medicine: Redmond and Gamlin review (1999) offers very broad and extensive list of the sensitizing structures with various quantum yields for the sensitization efficiency: chlorophylls, bacteriochlorophylls and their derivatives, as well as generally numerous types of porphyrin structures are among them and makes a significant part of the list.

1.2.3. Photosensitizers of 2nd And 3rd Generation: Chlorophylls and Bacteriochlorophylls

There is no precise classification what compounds could belong to a 1st generation of photosensitizers, but it has been accepted that second generation of photosensitizers for PDT induce light sensitivity in the region of 650-850 nm where most of the biological chromophores normally do not absorb light. The main 2nd generation photosensitizers are major photosynthetic pigments, chlorophylls and bacteriochlorophylls (Brandis *et al.* 2006a, 2006b). The Jablonsky diagram for chlorophyll molecule is shown at Figure 3-B. "Blue photon" absorption yields the higher, S_2 excited state, *i.e.* Soret-band, while the absorption in the red range yields S_1 (which also can be populated by non-radiation internal conversion – *ic*), *i.e.* Q_y-band. The S_1 state can be deactivated through a fluorescence emission, or can create the longer-lived, photochemically active T_1 state, responsible for Chl photosensitizing function. The Soret-band (Figure 2-B - also called "diagnostic" band) is not relevant for PDT, but the Q_y-band (also known as "curative band" - due to it's use in PDT) is responsible.

Unfortunately, these pigments are water insoluble which pose a ban for their use in PDT. However, since the photophysical characteristics of Chl and BChl are determined by their tetrapyrrole macrocycle, it was realized that modifications of peripheral groups that are not conjugated to the macrocycle may further enhance their hydrophilicity while retaining their photodynamic activity. That is a case of chlorins, pheophorbides and purpurins (Rosenbach-Belkin *et al.* 1998, and the references cited therein). Though quantum yield for T_1 state formation are somewhat lower compared to the native pigments, the new derivatives showed much higher photocytotoxicity (Rosenbach-Belkin *et al.* 1998, and the references cited therein). Following this strategy, the new Chl/BChl derivatives can be further conjugated with peptides, hormones or antibodies for targeting purposes (Scherz *et al.* 1994; Fiedor *et al.* 1996). That makes an introduction into 3rd generation photosensitizers.

Third generation photosensitizers are derivatives of 2nd generation photosensitizers, providing some biological specificity to deliver or to target such drug to cancer cells (Moser 1998). This includes coupling of photosensitizers with amino acids, polymers, proteins or carbohydrates. The porphyrin-protein complexes, especially those including various derivatives of Chl and BChl as a chromophores, have recently attracted a lot of attention (Schlichter *et al.* 2000; Proll *et al.* 2006; Markovic *et al.* 2007).

One of the major reasons for the development of the 3rd generation photosensitizers is a shift in the spectra of the 2nd generation photosensitizers in biological surroundings, due to aggregation and interaction with lipids or proteins. The shifts of more than 20 nm (of maximal absorbance value, A_{max}) are not uncommon. So, the spectral data obtained in organic solvents has little of real biological relevance (Moser 1998). In most cases absorption spectroscopy can not be performed in biological tissue (Moser *et al.* 1992), so some other optical techniques need to be applied to get some direct or indirect information about the status and properties of the photosensitizers inside the tumor environments. The fluorescence techniques is one of those used very often since it can reflect not only the status of the chromophore (Chl or BChl), but of it's biological matrix too (especially in the case of the pigment-protein complex, where fluorescence can originate not only from the chromophore, but from the protein amino acid residues, too - like tryptophan).

1.3. Chlorophylls in PDT Therapy – Overall

PDT therapy and use of chlorophyll and (generally) porphyrin type of sensitizers, is not unique, invariable method. Thanks to the variety of biological environments (cells, organs, tissues…) and related circumstances where it has been applied, PDT has been developed into few sub-methods based on the same theoretical principles. For example, photochemical internalization (PCI), a specific branch of PDT, is a novel technology utilized for the site-specific release of macromolecules within cells. The mechanism of PCI is based on the breakdown of the endosomal/lysosomal membranes by photoactivation of photosensitizers that localize on the membranes of these organelles (Berg *et al*. 1999). PCI has been shown to potentiate the biological activity of a large variety of macromolecules and other molecules that do not readily penetrate the plasma membrane (Berg *et al*. 2005). PCI method has been applied in gene therapy, a promising strategy to deliver desired gene into target cells for the treatment of genetic deficiencies (Shieh *et al*. 2008). There are few recently published reports on it, like the one related to gene expression (Fischer *et al.* 2007), or the ones where photosensitizers were part of a complex transporter delivered to the desired location, the most sensitive subcellar compartments (Rosenkrantz *et al.* 2003; Morgan *et al*. 2010). The sensitizers can also be "packed" in model systems, like liposomes, which then were used as a carriers to deliver a sensitizing drug to the appropriate target place: this is a case with metalloporphyrin and chlorophyll delivered from liposomes to inactivate methicillin-resistant *Staphylococcus aureus* (Ferro *et al.* 2006), though sometimes sensitizers presence is not necessary for drug delivery from the carrier, like with photo-triggerable DPPC-made vesicles for delivery of doxorubin to cells (Yavlovich *et al*. 2011). It is important to note that besides "pure" PDT (sensitizing) function chlorophylls and porphyrins may be used for fluorescence diagnosis (FD) of tumors as well (Berg *et al.* 2005; Juzeniene *et al.* 2007).

While photodynamic properties of Chl derivatives may be exploited *in vitro* (using human hepatocellular carcinoma cells – Li *et al*. 2007; for photoinactivation of vesicular stomatitis virus – Lim *et al*. 2002), PDT has already become widely used practical method that covers a number of medical fields including dermatology, gastroenterology, ophthalmology, blood sterilization and treatment of microbial-viral diseases. PDT can bring benefit to studies of cell organelles, mutagenic potential, tumors selectivity, cell cooperation (Juzeniene *et al.* 2007). List of potential PDT applications of (particularly) porphyrin photosensitizers (including chlorins, so chlorophylls, too) is given in Sternberg and Dolphyn review (1998); the cited treated diseases there range from eye macula (Age-related Macular Degeneration – AMD), *via* arthritis, arterial restenosis to prostate cancer. PDT treatment with particular Chl-derivative drug was used to study the inhibitory effects on Gross leukemia retrovirus isolated from mouse (Lee and Lee 1990).

1.3.1. PDT and The Skin

Skin "disorders" and diseases treated by PDT are of some enhanced interest for this chapter, and they include acne, psoriasis, Karposi's sarcoma and basal cell carcinoma (Sternberg and Dolphin 1998). It is important to note that today PDT-skin practice often uses porphyrin sensitizer precursor rather than porphyrins themselves. AminoLevulinic Acid (ALA) is very often externally added to the patients: it then passes the plasma membrane through amino acid transporters and enters the heme synthesis pathway leading to temporary accumulation of protoporphyrin IX (Berg *et al.* 2005; Sternberg and Dolphin 1998). Currently ALA is commercially licensed as Levulan, Kerastick (in USA), while methyl aminolevulate (commercially: MAL, Metvix) is licensed in Europe, Australia and USA (in the latter case only for treatment of Actinic Keratoses – AK). Unlike ALA which is hydrophilic and unstable, MAL has advantage of being more lipophyllic[5].

1.3.2. PDT, Skin Sensitivity and Chl Possible Actions

However, one common side effect of PDT skin therapy is skin photosensitivity. Because the free drug accumulates in the skin after treatment, some patients experience post-treatment skin sensitivity for as long as 6 weeks. Most of the drugs are activated by visible light, and therefore normal sunscreens preparations — which only guards against ultraviolet (UV) light — won't protect patients against rash and sunburns (depending on the application, light sources for PDT have included lasers, light emitting diodes (LED), and fluorescent light using delivery systems such as fiber optics, catheters, or endoscopes). The reported way to solve this problem is a development of a nanoparticle drug delivery system that may help eliminate skin photosensitivity. Ceramic-based nanoparticles—made with a class of

---

[5] From Clinical Photodynamics (Summer 2010 issue, Vol.2, No.5).

inert, nontoxic, and nonimmunogenic materials called organically modified silica—encapsulate and form a permeable membrane around the hydrophobic photosensitizers. This coating creates a water-compatible shell that enables the drug to be dispersed more readily and prevents its self-aggregation and loss of fluorescence. Once nanoencapsulated, the drug remains inside the particles. Because photosensitizers are not released from the particles, they are less likely to accumulate in the skin, thereby reducing phototoxic side effects[6].

Chlorophyll (or Chl-derivative) could actually might play a controversial role under particular circumstances: on one side it can work as photosensitizer (employed or not-employed in PDT therapy), which as a side consequence may lead to described skin sensitivity, up to the patients' pain feeling; on the other side chlorophyll is used as an active ingredient of some other – even patent protected (US Patent 7,700,079 B2; US Patent 7,201,765 B2) – dermatological applications for skin treatment from damaging effects caused by prolonged exposure to natural sunlight: there it plays few roles, pain-relief being just one of them (in this case it doesn't require a binding-ligand, as in PDT-case). Chlorophyll, as an active ingredient of the soap product promotes skin restoration, and treat ulcerative lesions and burns (US Patent 7,700,079 B2). It actually helps soothe sunburned skin – as a result of non-controlled exposure to natural sunlight – and provide pain relief, acting as a skin refresher. It has been also reported that chlorophyll accelerate wound healing and stimulate damaged tissue repair. Overall, chlorophyll promotes "wound healing, detoxify, deodorize and inhibits bacterial growth" (US Patent 7,700,079 B2).

However, and possibly in the same time, while acting on the (sunlight-induced) consequences (rashes, burns) – and if it is not part of PDT-skin drug and therefore it's sensitizing function has no any preference compared to the other options – Chl then might perform a preventive function, too, acting on a cause. The scheme of skin cross-section (Figure 4) shows a natural sunlight penetration through the skin: contrary to the energetic point of view, the shortest wavelength UV-B has the smallest and the longest-wavelength red light has the biggest penetration. The reason is in the presence of very strong UV-absorbers, endogenous chromophores, through various skin layers; there, their UV-absorption is mostly related to their sensitizing function. Wondrak and coworkers (2006) have made a list of the skin UV-photosensitizers: porphyrins, melanin, bilirubin, flavins, pterins, vitamin B_6, vitamin K,

---

[6] From: Kaylynn Chiarello: "*In between the light and the dark. Developments in photosensitive pharmaceuticals*", Pharmaceutical Technology, Special Report, December 2004, pp.52-54.

NAD(P)H etc (they may act both as Type I and Type II photosensitizers – Foote 1976, 1991; Girotti 2001), located inside particular type of cells (for example, melanin, responsible for the skin darkening as a result of prolonged natural sunlight exposure, is inside melanocytes cells). These chromophores make most of the skin damage to various targets inside it (like particular proteins, enzymes, DNA etc) if sunlight UV-component is not blocked by the (skin) applied sun-screen preparation. Still, they are week VIS light – especially red light absorbers - which may lead to other skin damage effects (burns, lesions etc, and, on the longer time scale to "skin aging"). Chlorophyll, the VIS light (esp. red light) efficient absorber - if used in the appropriate dermatological application - may absorb part of red light and therefore contribute to elimination of the other cited damage; as Figure 3-B shows (Chl Jablonsky diagram) the red light photon absorption may not necessarily lead to Chl sensitization function: the exciton may be released through the alternative pathways. So, in this case Chl could play prevention role.

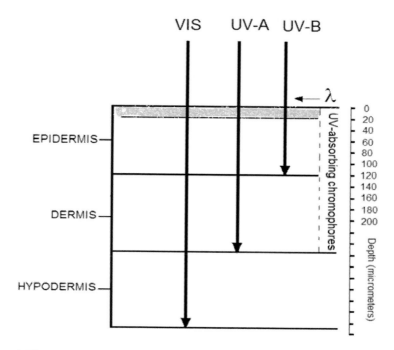

Figure 4. The presumed schematic penetration of UV-A and UV-B as well as VIS light into various depths of skin layers. The layers thickness is put a side. The endogenous chromophores (porphyrins and the derivatives being among the most abundant), located throught the layers, are most responsible for the attenuation of UV-A and UV-B penetration through the skin.

However, one should underline that chlorophyll healing and soothing actions help the skin to recover from a prolonged exposure to entire spectrum of natural sunlight, including UV-A and increasing portion of UV-B[7]. Therefore, in the frame of total response of Chl to natural sunlight the response to UV-portion emerges as a particular problem *per se*, and needs to be particularly elaborated.

2. Chlorophyll, Molecular Organization and UV-Effects

2.1. UV-Effects on Chlorophyll in a Broader Context

Solar radiation is the vital source of energy for the Earth biosphere. However, though being the ultimate driving force of photosynthesis and its important regulatory factor, solar light is also a major source of stress to photosynthetic organisms, *i.e.* it may act as an adverse environmental factor (see for example Holm-Hansen *et al.* 1993; Jansen *et al.* 1998). Generally, the efficiency of light-induced damage is proportional to the light energy. Therefore, from natural sunlight that reaches the Earth the UV-B (280-315 nm) component has the highest damaging potential. Thus, recent reductions in the stratospheric ozone layer, leading to enhancement of UV-B intensity on the surface of Earth and ecologically significant depths of the ocean[8] have multiplied extensive research efforts to elucidate molecular mechanisms by which various organisms respond to UV-B radiation (Teramura and Ziska 1996). As a major photosynthesis pigment chlorophyll is reportedly hit by UV-B action, and its composition is significantly altered when exposed to UV-B light *in vivo* and *in vitro* (Strid and Porra 1992). High levels of UV-B radiation in combination with low levels of Photosynthetically Active Radiation (PAR) have significantly reduced chlorophyll content in plants (Garrard *et al.* 1976;

---

[7] Depending on the sources reported, UV-portion of the total sunlight that reaches the Earth surface is 5-6%, and UV-A (320-400 nm) makes >95% of it, under normal circumstances; The UVB (290–320 nm) content of total solar UV-flux on skin can be well below 2% depending on the solar angle, which determines the atmospheric light path length and thereby the degree of ozone-filtering and preferential Rayleigh scattering of short wavelength UV light (Wondrak *et al.* 2006).

[8] "The recent reductions" of ozone layer (and the related increase UV-B intensity on the Earth surface) are time (season)- and space-variable categories and depend on many factors, like latitude and altitude.

Tevini *et al.* 1981; Vu *et al.* 1984). As a consequence of this reducement, and not surprisingly having in mind central Chl role in photosynthesis, serious misfunctions and damages related to the particular entities of photosynthetic apparatus have been reported as a result of not only UV-B (Vass 1997; Renger *et al.* 1989; Larkum *et al.* 2001) but of UV-A action as well (Turcsányi and Vass 2000; Vass *et al.* 2002). While these particular (UV-A and UV-B *vs* Chl-photosynthesis-functioning) effects (Bornman 1989) are not in the focus of this contribution they should not be neglected, because they implicitly suggest at least two crucial things. First, having in mind very complex structure of photosynthetic apparatus and it's belonging major entities (two large light-harvesting pigment-protein complexes, known as PSII and PSI, embedded in lipo-protein matrix of thylakoid membranes – Whitmarsh and Govindjee 1995) it suggests that Chl stability to UV-radiation is very probably related to it's molecular environment and related organization (Zvezdanović and Marković 2008; Zvezdanović *et al.* 2009). Second, as reported, the mechanism of Chl-UV interaction is not simple, but complex: it most often includes intermediates and very reactive free radicals and other oxygen (ROS) species (Faller *et al.* 2001; Hideg *et al.* 1998, 1999). Exposure of Chls to UV-A and UV-B *in vivo* is believed to enhance the amount of ROS (Barta *et al.* 2004; Zhang *et al.* 2005). Hydroxyl radicals have been pointed out to take participation in Chl photooxidation in aqueous micellar systems (Harbour and Bolton 1978; Chauvet *et al.* 1981).

One should add at this place that generally studies of UV-irradiation on Chl structure and function are pretty rare. They are usually reported in the frame of studies dealing with Chl response towards external radiation (or: Chl stability toward external radiation stress), which are in the most cases related to the visible light induced-effects, rather than on UV-induced effects. Sometimes these studies are in the broader and more general frame of chemistry of chlorophyll modifications (Inhoffen 1968; Hynninen 1991). Though these (also rare) studies are generally known as "photooxidation studies", if not otherwise indicated they deal primarily with VIS-light induced effects. Even started long time ago (see for example, Aronoff and MacKinney 1943; Woodward 1961), and despite a huge innovations in modern analytical techniques over the last two-three decades, the structure of the obtained photoproducts is relatively unknown yet, and the mechanisms by which they have been obtained are still very disputable or simply unknown – as it had been cited 30 and 40 years ago (when the intermediates and products are simply identified by colour, like "red" or "reddish" – Livingston and Stockman 1962; Morris *et al.* 1973; Jen and MacKinney 1970). Part of explanation for this enigma is in the undoubtfull fact that chlorophyll is not an efficient UV-absorber but still is able to absorb UV-radiation (Johnson and Day 2002); another fact is in undisputed complexity of the process. In the very extensive review Hendry *et al.* (1987) put photochemical oxidation of chlorophyll in the much broader frame of "chlorophyll degradation" or "breakdown", induced by various external or internal factors like senescence (part of Chl life cycle), metabolism, temperature, pollutants etc; a support for this, much broader Chl breakdown picture might come from the fact that some very same Chl degradation products obtained by different degradation processes (like senescence and photooxidation) have been detected (Hendry *et al.* 1987), obviously from very different "Chl-sources" with belonging molecular organizations[9].

2.2. In Vitro Studies

The studies *in vitro* deal mostly with chlorophylls (different forms of Chl, or mixture of Chls) in solution, but that does not mean implicitely a homogeneous environment; that is the case for example with photooxidation studies of chlorophylls in detergent solutions (Chauvet *et al.* 1973; Shaposhnikova *et al.* 1973). Chlorophyll is capable of building aggregates *in vitro* (as well as *in situ*), which is based on it's electron donor-acceptor characteristics (Katz *et al.* 1963; Katz *et al.* 1978). In chlorophyll, central magnesium (Mg) coordinates four nitrogen atoms of the pyrrole rings providing one or both Mg axial positions to be occupied by a (donor) molecule having alone electrons pair. When Chl is dissolved in a nucleophilic polar solvent such as acetone, the solvent acts as electron donor to Mg, and chlorophyll then appears as "monomer", with five- and/or six-coordinated Mg (Katz *et al.* 1978). On the other hand, Chl may act as electron donor, too, due to presence of few keto groups, mostly belonging to the peripheral ester groups (C=O group in position C-13^1 in Chl*a* molecule, additional C=O group in Chl*b* molecule in position C-7, and two ester groups in positions C-13^3 and C-17^3, both in Chl*a* and Chl*b* – Figure 1).

Of course, Chl response to VIS or UV radiation *in vitro* is not uniform and depends on at least few factors; the most important among them are: presence

---

[9]The studies on bacteriochlorophylls (BChls) photooxidation are also rare (Lindsay-Smith and Calvin 1966; Troxler *et al.* 1980; Brown *et al.* 1980; Steiner *et al.* 1983) and mainly face the same difficulties as in the case of Chl photooxidation.

(or absence) of oxygen, radiation regime, and already mentioned possible aggregation. What any of these (and other additional) factors don't put in question is degradation, or decomposition, or bleaching of chlorophyll that takes place. The VIS-induced bleaching could be reversible in different solvents under anaerobic conditions (as a "side effect" this report cites an interesting fact: the bleaching of demetalated Chls, *i.e.* pheophytins, are scarcely detectable - Livingston and Stockman 1962). Under aerated conditions the irreversible bleaching dominates in solution, and it is accompanied by formation of reddish or pink intermediates, precipitated by addition of highly nonpolar solvents in appropriate solvent mixtures (Jen and MacKinney 1970; Morris *et al.* 1973); at the end a colorless photodegradation product has been formed (Llewellyn *et al.* 1990a). Some of (VIS-induced) pheophytin oxidation products in aqueous suspension have been detected by GC-MS analysis, and they include some simple, non-cyclic compounds like lactic acid, oxalic acid, citric acid, glycerol, and some cyclic peroxide structures, too (Llewellyn *et al.* 1990b). The presence of the simple products certainly presumes an opening of Chl macrocyclic ring; Struck and collaborators prove it (1990) by continuous VIS-light irradiation of 20 chloro-Chls (substituted in position 20 – Figure 1): they've found as the intermediates long-wavelength shifted photoproducts, which have been transformed in the dark afterwards to linear, non-cycling (or: opened-chain) tetrapyrroles (bilin pigments).

A rare try to elaborate photooxidation mechanisms of Chls came from Rontani and his group. In a series of papers they've reported that in the presence of visible light and oxygen Chl photooxidation may be "structurally divided" to the porphyrin ring oxidation and the phytol chain oxidation; especially in the latter case they pointed out photosensitizing Chl-triplet (^{3}Chl) as a major role player, since it creates singlet oxygen that attacks phytol chain double bonds and leads to a lot of isoprenoid product compounds (Rontani *et al.* 1991; 1995; Cuny and Rontany 1999; Cuny *et al.* 1999). Here, chlorophyll don't act only as an oxidation agent (through photosensitization) but undergoes to oxidation itself ("autooxidation").

The UV-induced bleaching of chlorophylls, didn't change basic conclusion about the irreversibility effect as the final outcome obtained under aerobic conditions (seen with VIS-induced bleaching), even when the applied light regime has been changed from continuous irradiation to the excitation flashes: (in the latter case) the rate of UV-induced bleaching of bacteriochlorophylls and chlorophylls rises with increase of oxygen concentration (Fiedor *et al.* 2002; Drzewiecka-Matuszek *et al.* 2005, respectively). However, the extent of the bleaching will be highly depended on the molecular organization of Chl molecules in the particular solvent. In the recently published reports (Zvezdanović and Marković 2008; Zvezdanović et al. 2009) we've compared UV-induced bleaching of chlorophylls (from extract of photosynthetic pigments isolated from spinach leaves – so containing both Chl*a* and Chl*b*) in acetone and *n*-hexane, where two different forms of Chl organization have been found. In a nucleophilic polar solvent such as acetone, the solvent acts as electron donor to Mg, and chlorophyll then appears as "monomer". On the other hand, in nonpolar solvents such as *n*-hexane, chlorophyll predominantly appears in aggregated forms ("dimers" and "oligomers") at a higher concentrations (over the 10^{-5} moldm^{-3}) (Katz et al.1978; Oba et al.1997), and as a "monomer" at a lower concentration (Katz et al.1978); there, one chlorophyll may act as an electron donor (*via* its ring E, keto group, C-13^1 position, Figure 1) and the other Chl molecule can act an electron acceptor *via* its` central Mg atom (Trifunac and Katz 1974). The Q_y-band A_{max} of "monomeric" Chl in *n*-hexane lies at ~660 nm (at $c_{(Chla+Chlb)}=1.2\times10^{-6}$ moldm^{-3}). In concentration ranges about 10^{-5} and 10^{-4} moldm^{-3}, Q_y-band A_{max} is "red" shifted due to Chl-Chl interaction in "dimeric" chlorophyll form (λ_{max}~665 nm) (Zvezdanović and Marković 2008; Zvezdanović et al. 2009)[10].

Bleaching of the extracted Chls in *n*-hexane in "dimeric form" is shown in the VIS absorption spectra, Figure 5-A. A hypochromic effect has been seen both for Q_y and Soret-band, and the belonging kinetics for all three used UV-ranges (UV-A, UV-B and UV-C), reflecting decrease of two bands absorption maxima (at 430 and 665 nm, respectively) is shown in Figure 5-B; the kinetics clearly obey to 1st order, with the biggest rate constants obtained with UV-C (compared to UV-A and UV-B), indicating the incident UV-photons energy to play major role in the "quantity" of UV-induced bleaching. The same linear 1st order plot of degradation was found for UV-irradiated Chls in postharvested fruit and vegetables (Huang et al. 2008). The existence of the two intersection points (at ~487 and ~565 nm) at Figure 5-A, in the range of 450-600 nm where a small, but detectable increase of absorption has been observed (Zvezdanović and Marković 2008; Zvezdanović et al. 2009; White and Tollin 1971; Merzlyak et al. 1996), suggests indirectly a possible formation of intermediate products (or *transients*), as a consequence of UV-induced Chl degradation. The transients formation is further supported by

---

[10] A natural model for Chl "dimer" exists in anoxygenic photosynthetic bacteria: a bacteriochlorophyll, BChl-dimer in reaction center (Allen and Williams 2006).

recording the fluorescence spectra of the Chl-fraction and spinach leaves extract (containing not only Chls but carotenoids, too!) in *n*-hexane in the same 400-800 nm range, following UV-continuous irradiation from the three UV sub-ranges (UV-A, UV-B and UV-C) (Zvezdanović *et al.* 2009): clear rise of fluorescence emission in the 450-500 nm range (known also as blue-green range, so naming this emission as Blue-Green Fluorescence, or BGF emission) – found both with Chl-fraction as well as with the extract – and synchronous with decrease of Q_y-band maximum fluorescence intensity ($F_{max} \sim 670$ nm) has been detected – Figure 6-A. The corresponding, rising kinetics plots for the transients formation in BGF range (Figure 6-B), obeying again to 1st order suggests that the transients origin might be related to Chl bleaching. Merzlyak and collaborators (1996) who irradiated Chl solution and the extract of photosynthesis pigments with continuous both VIS and UV-light have also come up to this conclusion; to make the case stronger they irradiated carotenoids fractions only under the same conditions (as Chls) and recorded the spectra in 350-650 nm range afterwards: the absence of visible changes strengthened the presumption of Chl-origin of the transients. Moreover, there are few reports concerning Chl photooxidation in solutions, where an opening of the Chl porphyrin ring was suggested as an initial step for the transients structures formation (Jen and MacKinney 1970; Brown *et al.* 1980; Hendry 1987; Struck *et al.* 1990; Brown *et al.* 1991).

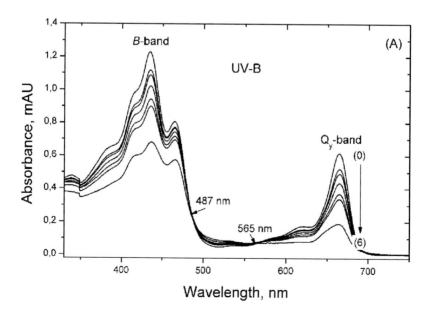

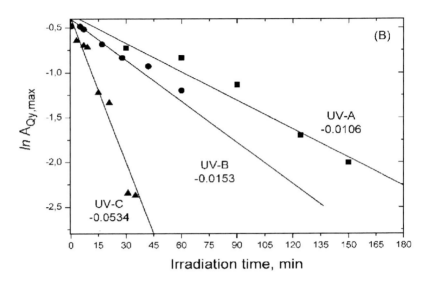

Figure 5. Chlorophyll bleaching in pigments extract (isolated from spinach leaves) in *n*-hexane, induced by continuous UV-B-irradiation. The extract contains > 80 % of Chl*a* and Chl*b* (A) Changes in the extract absorption spectra with the increasing irradiation periods: (0) 0 min; (1) 5 min; (2) 7 min; (3) 17 min; (4) 28 min; (5) 30 min; (6) 42 min. (B) Kinetics of the bleaching shown under (A), as well as of UV-A and UV-C induced bleaching (the sample from the same extract bulk), obtained by tracing maximum absorption values of Chl Q_y-band (A_{max} of Q_y-band) as a function of UV-irradiation periods ($ln\ A_{Qy,max} = f(t_{irr})$). The corresponding bleaching rate constants (in min^{-1}) have been shown on the plot. Chl concentrations are: $c_{Chla}=8{,}5\cdot10^{-6}$ moldm^{-3}, $c_{Chla+Chlb}=1{,}1\cdot10^{-5}$ moldm^{-3}. (*From*: Zvezdanović *et al.* 2009).

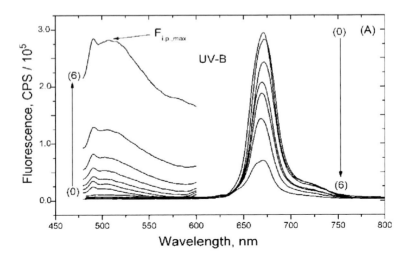

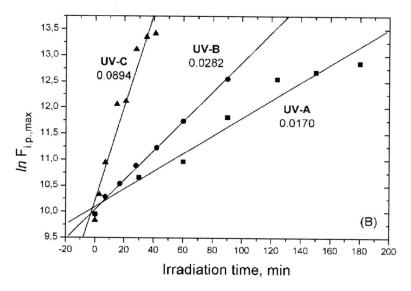

Figure 6. (A) Changes in fluorescent spectra of the pigments extract (isolated from spinach leaves) in *n*-hexane, induced by continuous UV-B irradiation, detected in Blue-green, red, and far red range (BGF, RF and FRF), respectively. The increasing UV-B irradiation periods are (0) 0 min; (1) 7 min (2) 17 min; (3) 28 min; (4) 42 min; (5) 60 min; (6) 90 min. (B) Kinetics of formation of the intermediare products (*the transients*), followed through a rise of maximum fluorescence emission in BGF range ($F_{i.p.,max}$) with the increased lengths of UV-B (synchronous with Chl-bleaching shown under (A)), as well as of UV-A and UV-C irradiation periods (the sample from the same extract bulk): $ln\ F_{i.p.,max} = f\ (t_{irr.})$. The corresponding calculated transients formation rate constants ($k_{i.p.}$, min^{-1}) have been shown on the plot, for all three UV sub-ranges. Chl concentrations are: $c_{Chla}=8,5·10^{-6} moldm^{-3}$, $c_{Chla+Chlb}=1,1·10^{-5} moldm^{-3}$. (*From*: Zvezdanović *et al*. 2009).

Finally, it is interesting to note that this type of the (UV-induced) Chl spectral changes (shown at Figure 5-A and Figure 6-A) has been seen not only with continuous irradiation of Chl solution with VIS and UV-light (Merzlyak *et al*. 1996; Zvezdanović and Marković 2008; Zvezdanović *et al*. 2009), but with laser 355 nm flashes as well (Drzewiecka-Matuszek *et al*. 2005).

In polar acetone, the similar type of response (seen as the one described in the upper chapter in *n*-hexane) has been seen – but only speaking in qualitative manner: the bleaching rate constants, and the quantity of bleaching, taken overall appeared to be significantly different (Zvezdanović and Marković 2008; Zvezdanović *et al*. 2009). Due to acetone' carbonyl "bridges", electron density of chlorine ring is not arranged between two chlorophyll molecules (like in Chl "dimer") but is rather distributed between one chlorophyll molecule and one or two acetone molecules (Katz *et al.* 1978). Since acetone absorb in whole UV range (Yujing and Melluki 2000), that may lead to a declined stability of the "monomer" Chls against UV-radiation, compared to "dimer" or "oligomer" Chls in *n*-hexane. Comparison of photobleaching rate constants, as well as the transients formation rate constants for the three UV sub-ranges obtained in the two solvents, supports this statement (Table I – Zvezdanović *et al.* 2009); for all three UV sub-ranges the bleaching rate constants as well as the corresponding transients formation rate constants are much higher in acetone than in *n*-hexane, indicating higher sensitivity of "monomer" Chls compared to the "oligomer" ones. So, based on these data one may conclude that "dimer" or "oligomer" molecular organization of Chls in nonpolar *n*-hexane have protected them against UV-irradiation, compared to their "monomer" configuration in highly polar acetone.

Table I. Chlorophyll bleaching and the transient formation kinetics for the pigment extract chlorophylls in acetone and *n*-hexane, and in aqueous suspensions of thylakoids, during increasing UV-irradiation intervals in three different UV-ranges: UV-A, UV-B and UV-C. Chlorophyll absorbances were followed at Amax Q-band values (662 nm in acetone, 665 nm in *n*-hexane and 679 nm in thylakoids). The transient formation was followed at $F_{trans.,max}$ values: ~515 nm in acetone, ~510 nm in *n*-hexane for all three UV-ranges; ~515 nm for UV-A and -B induced transient formation, and at ~550 nm for UV-C transient formation, in thylakoids suspensions. Rate constants (k), k_{Chl} and $k_{trans.}$ are calculated in min^{-1}.
(From: Zvezdanović et al. 2009)

	Extracted chlorophylls in acetone		Extracted chlorophylls in *n*-hexane		Chlorophylls in thylakoids	
	k (min^{-1})					
UV-irradiation wavelength (nm)	Chl-bleaching	Transient formation	Chl-bleaching	Transient formation	Chl-bleaching	Transient formation
UV-A (350)	-0.0859	0.1797	-0.0106	0.0170	-0.0100	0.0064
UV-B (300)	-0.2399	0.2281	-0.0153	0.0282	-0.0145	0.0205
UV-C (254)	-0.2260	0.2031	-0.0534	0.0894	-0.0654	0.1121

2.3. In Situ Studies

This fact becomes even more clear with more organized Chls, the ones that are aggregated in the antennas of the pigment-protein complexes (photosystem II and I: PSII and PSI, respectively) (Green et al. 2003; Parson and Nagarajan 2003), with the main function to absorb and redistribute the incident photons towards the reaction centers (Figure 7)[11]. While any of Chl roles in photosynthesis is not of particular interest for this chapter, it is necessary to add that high efficiency of the (incident) photons transfer inside the antennas is provided by the supramolecular organization, where Chls molecules are tightly packed inside lipo-protein matrix, permitting overlapping of their absorption spectra and therefore an excellent photon transfer from one Chl molecule (donor) to another one (acceptor) (van Amerongen and Dekker 2003; Parson and Nagarajan 2003)[12]. That is why isolated photosynthetic organelles, chloroplasts, and sub-organelles, isolated thylakoids (which are made by chloroplasts break-up – Cerovic and Plesnicar 1984; Walker et al. 1987) make an excellent natural example of superior Chl organization, which – not surprisingly – can serve as a practical *in situ* model for studies of VIS and UV-light influence on Chl structure and function (Carpentier and Leblanc 1986; Merzlyak 1989; Krasnovsky et al. 1994; Santabarbara 2006). Of course, since the investigated system is now more complex than solution, the mechanisms and the pathways of Chl degradation are more complicated, and the structure of products is largely unknown – the only certain thing is that these complicated photooxidative transformations are mediated by various forms of activated oxygen (Hendry 1987; Brown 1991; Merzlyak 1989; Merzlyak et al. 1991; Krasnovsky Jr 1994; Merzlyak and Hendry 1994); if the products are fluorescent they certainly have Chl-origin (Karuktsis 1991). Of course, under *in vivo* conditions, when the object of UV-induced action are whole organisms, like phytoplankton (see reviews like Vincent and Neale 2000; Banaszak and Neale 2000), or whole plants (Jansen et al. 1998), or whole photosynthesis process (Neale et al. 1993; Teramura and Ziska 1996), or some particular phase of it (Melis et al. 1992; Bouchard et al. 2005; Hakala et al. 2005), the eventual effects on chlorophyll structure and functions are just

---

[11] ...where the primary photochemical act, a charge separation takes place, but this is not a focus of this contribution so will not be further elaborated.

[12] The supramolecular organization does not contain just Chls, but carotenoids (accessory photosynthesis pigments) as well (Green et al.2003); their main function is to absorb the photons in between "blue" and "red" range of the sunlight spectrum (where Chl absorption reach maximums), so providing sunlight absorption at all belonging wavelengths and therefore providing optimal conditions for this step of photosynthesis light phase.

one of the consequences and can not be separated from the other *in vivo* processes, but this is not the subject of this chapter.

How the stability of Chls is affected by UV-irradiation in isolated thylakoids supramolecular organization, compared to solution environment – what should be expected? As it is well known, during photosynthesis process in healthy chloroplasts and thylakoids, chlorophyll is protected from light destruction by surrounding carotenoids and lipids, and it's high-ordered molecular organization. However, when such protection is lost chlorophyll becomes vulnerable in the presence of light and oxygen[13].

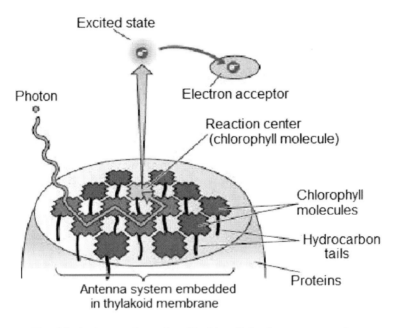

Figure 7. Simplified scheme of transfer of incident light photons among the aggregated Chl molecules inside the antennas of isolated thylakoids (carotenoids molecules are not shown).

---

[13] The other main carotenoids function in the antenna of PSII and PSI – besides absorption of light in mid-region (roughly, between 450-600 nm, where Chls do not absorb) and it's transfer to the neighboring Chls – is photoprotective function: it is known that the excess of light that is not needed to run photosynthesis (so, the over-PAR intensities) – that could potentially be very harmful to Chl structure and function – is used by carotenoids in a cycle of chemical reactions, where epoxy-type of structures have been synthesized (see Peter Horton group papers about it, like Pascal *et al.* 2005; Ruban *et al.* 2007); that is why one should expect higher resistance, or stability of chlorophyll toward light – as long as the thylakoids structures is kept intact – compared to Chls in solution.

One should say again that – as well as in the case with Chls in solution – the studies dealing with Chls response to VIS and UV-light inside thylakoids or their sub-particles are even more rare than in the case of the solution-related studies (Merzlyak *et al.* 1996; Cerovic *et* al. 1999; Latouche *et al.* 2000; Olszowka *et al.* 2003; Santabarbara 2006; Hirabayashi *et al.* 2006; Zvezdanović *et al.* 2009) – unless the emphasis is put on a particular photosynthesis related processes, where the influence (of VIS or UV-light) on Chls is considered as an additional, but minor effect; in such a case – which again is not focus of this contibution – there are much more citations (see pp.16-17). One could say that these results, though obtained by use of different spectroscopies – *i.e.* absorption VIS and fluorescence spectroscopy (Merzyak *et al.* 1996; Zvezdanović *et al.* 2009) or with VIS and CD (circular dichroism) spectroscopy (Santabarbara 2006) – express some consistency. Generally, in all cases Chl bleaching has been confirmed – seen through hypochromic effect of Chl main absorption / fluorescent band and the corresponding increase of absorption / fluorescence emission in 450-650 nm range with the increase of the irradiation periods, which is indication for the intermediate products formation (Merzlyak *et al.* 1996; Santabarbara 2006, Zvezdanović *et al.* 2009). Inside this very broad frame, there are specificities depending on whether one deals with VIS or UV-effects (and inside UV-range, UV-A, UV-B or UV-C), on chloroplasts, thylakoids, or on it's sub-particles; sometimes, the effects are simply better seen with one techniques rather than with the other one. For example, by doing experiments on isolated thylakoids Santabarbara (2006) showed that a continuous irradiation by VIS light leads not only to decrease of Q_y A_{max} values (~ 680 nm) - *i.e.* hypochromic effect, which is kind of general response - but to a hypsochromic effect as well. This "blue shift" has been attributed to a slower oxidation of Chl*b*, which Q_y A_{max} value is always "blue-shifted" compared to Chl*a* (in thylakoids, Q_y A_{max} position for Chl*b* is at 652 nm, compared to about 668 nm for Chl*a*) and therefore causes a delayed appearance of left shoulder on Q_y band. But maybe the most detailed response of Chls inside thylakoids was obtained by using fluorescence, because Chl fluorescence emission from thylakoids (and from whole intact photosynthetic organisms like plants, algae, bacteria) has long time ago been established as an intrinsic probe that monitors various stages of photosynthesis process (Karukstis 1991; Govindjee 2004; Nedbal and Koblizek 2006).[14] Fluorescence

---

[14] But, making this statement more relative, one should take care about possible Chl "auto-absorption": Chl may absorb fluorescence emission that has been created by its original absorption – this phenomenon is more or less expressed depending on the involving medium (Sauer and Debreczeny 1996).

spectra of UV-irradiated Chls cover three ranges (both *in vitro* and *in situ*): already mentioned blue-green range (*Blue-Green Fluorescence*, BGF, from 400-600 nm), red range and far-red range (*Red Fluorescence*, RF, from 630 – 700 nm, and *Far Red Fluorescence*, FRF, from 700-800 nm, respectively); in the case of Chls in thylakoids and chloroplasts, F_{max} in RF range lies at 681 nm and comes from Chl*a* from PSII antenna, while a shoulder at 735 nm in FRF has origin from both PSII and PSI antennas (Cerovic *et al.* 1991). On the other hand BGF emission is assigned to Chl degradation product (Latouche *et al.* 2000) – that is why F_{max} values for RF Q_y band (Zvezdanović *et al.* 2009), as well as for RF and FRF Q_y band (Cerovic *et al.* 1999) suffered hypochromic effect, while BGF emission showed almost linear rise with the increased length of the irradiation periods, following order UV-C > UV-B > UV-A (Zvezdanović *et al.* 2009) – Figure 6-B.

Finally when the Chl bleaching rate constants as well as the transients formation rate constants (calculated from fluorescence measurements) have been mutually compared for "monomeric" Chls (acetone), "oligomeric" (*n*-hexane) and "aggregated" Chls (in thylakoids - Table I) – when these rate constants have been obtained under the same experimental conditions (the same irradiation regime with UV-A, UV-B and UV-C lamps) – some relationship between chlorophyll stability and degree of molecular organization of the involving surroundings has emerged (Table I – Zvezdanović *et al.* 2009): for all three UV sub-ranges there is clear distinction between "monomer" Chls in acetone on one side, and "oligomeric" and "aggregated" Chls (in *n*-hexane and thylakoids, respectively) on the other side – "monomeric" Chls are easily bleached compared to the latter, "organized" ones, which bleaching had pretty close rate constants (and in both cases "jump" in bleaching rates has been found with UV-C, compared to UV-A and UV-B, while with "monomeric" Chls UV-B made the difference). About the same type of relationship (between "monomeric" and "organized" Chls) was found by comparison of the corresponding transients formation rate constants (Table I). The additional weight to this conclusion comes from the fact that the bleaching rate constants have been calculated from VIS absorption data (*i.e.* A_{max} Q_y-band values have plotted against the irradiation time periods), while the transients formation rate constants have been calculated from the plots obtained from fluorescence data (*i.e.* F_{max} values corresponding to the particular fluorescence transients found in the 450-650 nm range). It looks that Chls bleaching and the transients formation are mutually connected, and even drastic change in Chls molecular organization (from "monomeric" to "oligomeric"), or in surrounding environment (complex organization of isolated thylakoids, involving – besides aggregated Chls – carotenoids, metalloproteins, enzymes, ...all attached on lipo-protein matrix) does not seem to put in question this fact. In other words, it seems that a substantial change in molecular organization and surroundings has changed Chl response to UV-irradiation more in quantitative than in qualitative manner.

Conclusion - The Biomedical Impact

What could be the "biomedical impact" of Chl (un)stability effects induced by UV-irradiation and it's dependence on molecular organization? It mostly concerns use of chlorophyll for various dermatological applications to treat the skin subjected to a prolonged natural sunlight exposure, since PDT therapy with Chl needs VIS and not UV-light, and therefore is not at least directly related to this stuff, while some indirect implications can not be neglected (having in mind the reported similar consequences that were found in the experiments with VIS and UV-light). UV-A and increasing portion of UV-B are parts of natural sunlight spectrum that touches the skin, and that is why – to avoid especially UV-B damaging action - the UV-absorbing compounds (besides light-reflecting compounds, like titan-dioxide) are often ingredients of the "normal" sun-screen preparations (protecting the skin from UV). As a week UV-absorber chlorophyll does not perform this function, but it is used in the other dermatological formulations as a skin restorer, refresher (to heal rash and lesions produced by natural sunlight), and a pain reliever. In the latter case it actually might play a double role: damaging through photosensitization (by sunlight absorption from "red" range, *via* it's triplet) that may lead to a variety of (photo)oxidized products, and healing function, acting on burns and sub-burns, produced by a prolonged skin exposure to natural sunlight. In both cases chlorophyll is affected by UV-portion of sunlight spectrum, and it's stability depends on surrounding molecular organization. Generally, if the environment forces a tight packing, chlorophyll should be more protected since it exists in more or less aggregated form. Because the dermatological applications very often use (as a raw material) different organized, lipoidal forms (liposomes, vesicles etc) in the frame of oil-water emulsions for the final preparations, it is reasonable to expect that chlorophyll, as a possible ingredient of such formulations, could have some protection from UV-irradiation – to some extent: one should not neglect numerous and various UV-irradiation effects on the environment, like lipid peroxidation[15] – just to mention one of these – which in turn may affect stability of chlorophyll itself (especially in the case of free radical LP mechanism (Foote 1976; Girotti 2001), involving a lot of ROS "attacking" species). Still, despite the controversies about it's effective role and the lack of an entire picture of UV-Chl interactions under these circumstances, the UV-Chl studies are worth to work on and a pretty limited number of published reports should rise, with an obvious impact on biomedical applications, concerning primarily the methods where the light is used as a sensitizing agent.

Acknowledgments

This chapter was supported under the Project on Development of Technology number TR-34012 by the Ministry of Education and Science of the Republic of Serbia.

References

Allen, J.P., Williams, J.C. (2006). The influence of protein interactions on the properties of the bacteriophyll dimer in reaction centers. In B. Grimm, R.J. Porra, W. Rüdiger and H. Scheer (Eds.) *Advances in Photosynthesis and Respiration: Chlorophylls and Bacteriochlorophylls* (pp.283-295). Dordrecht, The Netherlands: Springer.

Aronoff, S. and MacKinney, G. (1943). Photo-oxidation od chlorophyll. *Journal of the American Chemical Society, 65*, 956-958.

Banaszak, A.T. and Neale, P.J. (2001). Ultraviolet radiation sensitivity of photosynthesis in phytoplankton from an estuarine environment. *Limnology and Oceanography, 46*, 592-603.

---

[15] Lipid peroxidation (LP), very broadly defined as an oxidation process of membrane lipids (which can be triggered by various mechanisms, where the input of an external radiation - like UV - appears as just one of at least few options) , is one of the most general consequences of UV-action *in vivo* (membranes, cells, tissues, organs) as well as *in vitro* ("biological models"); the readers are recommended to look at an excellent Girotti (2001) review, to get more knowledge about LP mechanisms, reaction pathways, cytotoxical effects etc.

Barta, C., Kalai, T., Hideg, K., Vass, I. and Hideg, E. (2004). Differences in the ROS-generating efficacy of various ultraviolet wavelengths in detached spinach leaves. *Functional Plant Biology, 31*, 23-28.

Berg, K., Selbo, P.K., Prasmickaite, L., Tjelle, T.E., Sandvig, K., Moan, D., Gaudernack, G., Fodstad, O., Kjolsrud, S., Anholt, H., Rodal, G.H., Rodal, S.K. and Hogset, A. (1999). Photochemical internalization: a novel technology for delivery of macromolecules into cytosol. *Cancer Research, 59*, 1180–1183.

Berg, K., Selbo, P.K., Weyergang, A., Dietze, A., Prasmickaite, L., Bonsted, A., Engesaeter, B.Ø., Angellpetersen, E., Warloe, T., Frandsen, N. and Høgset, A. (2005). Porphyrin-related photosensitizers for cancer imaging and therapeutic applications. *Journal of Microscopy, 218*, 133–147.

Bornman, J.F. (1989). Target sites of UV-B radiation in photosynthesis of higher plants. *Journal of Photochemistry and Photobiology B: Biology, 4*, 145-158.

Bouchard, J.N., Campbell, D.A. and Roy, S.(2005). Effects of UV-B radiation on the D1 protein repair cycle of natural phytoplankton communities from three latitudes (Canada, Brazil and Argentina). *Journal of Phycology, 41*, 273-286.

Brandis, A.S., Salomon, Y. and Scherz, A. (2006a). Chlorophyll sensitizers in photodynamic therapy. In B. Grimm, R.J. Porra, W. Rüdier and H. Scheer (Eds.), *Advances in photosynthesis and respiration: Chlorophylls and Bacteriochlorophylls* (pp. 461-483). Dordrecht, The Netherlands: Springer.

Brandis, A.S., Salomon, Y. and Scherz, A. (2006b). Bacteriochlorophyll sensitizers in photodynamic therapy, In B. Grimm, R.J. Porra, W. Rüdier and H. Scheer (Eds.), *Advances in photosynthesis and respiration: Chlorophylls and Bacteriochlorophylls* (pp. 485-494). Dordrecht, The Netherlands: Springer.

Brown, S.B., Houghton, J.D. and Hendry, G.A.F. (1991). Chlorophyll Breakdown. In H.Scheer (Ed.) *Chlorophylls* (pp.465-489). Boca Raton, FL: CRC Press.

Brown, S.B., Smith, K.M., Bisset, G.M.F. and Toxler, R.F. (1980) . Mechanism of photooxidation of bacetriophyll derivatives: a possible model for natural chlorophyll breakdown. *Journal of Biological Chemistry, 255*, 8063-8068.

Carpentier, R. and Leblanc, R.M. (1986). "Chlorophyll photobleaching in pigment-protein complexes. *Zeitschrift fur Naturforschung C, 41C*, 284-290.

Cerovic, G., Samson, G., Morales, F., Tremblay, N. and Moya, I. (1999). Ultraviolet-induced fluorescence for plant monitoring: present state and prospects. *Agronomy, 19*, 543-578.

Cerović, Z.G. and Plesnicar, M. (1984). An improved procedure for the isolation of intact chloroplasts of high photosynthetic capacity. *Biochemical Journal, 223*, 543-545.

Chauvet, J-P., Journeaux, R. and Viovy, R. (1973). Photo-oxydation de la chlorophylle a dans le binaire Triton X-100–eau. *Comptes Rendus de l'Académie des Sciences C. 277*, 527–530.

Chauvet, J-P., Villain, F. and Viovy, R. (1981). Photooxidation of chlorophyll and pheophytion. Quenching of singlet oxygen and influence of the micellar structure. *Photochemistry and Photobiology, 34*, 557-565.

Chiarello, K. (2004). In between the light and the dark. Developments in photosensitive pharmaceuticals. *Pharmaceutical Technology, Special Report*, pp.52-54.

Cho, S., Lee, D.H., Won, C-H., Kim, S.M., Lee, S., Lee, M-J. and Chung, J.H. (2006). Drink containing chlorophyll extracts improves signs of photoaging and increases type I of procollagen in human skin in vivo. *Korean Journal of Investigative Dermatology, 13*, 111-119.

Cuny P., Romano, J-C., Beker, B. and Rontani, J-F. (1999). Comparison of the photodegradation rates of chlorophyll chlorin ring and phytol side chain in phytodetritus: is the phytyldiol versus phytol ratio (CPPI) a new biogeochemical index?. *Journal of Experimental Marine Biology and Ecology, 237*, 271–290.

Cuny, P. and Rontani, J-F. (1999). On the widespread occurrence of 3-methylidene-7,11,15-trimethylhexadecan-1,2-diol in the marine environment: a specific isoprenoid marker of chlorophyll photodegradation. *Marine Chemistry, 65*, 155-165.

Doss, J.C. (2010). Therapeut soap product with UV protection. *US Patent 7,700,079 B2*.

Dougherty, T.J., Gomer, C.J., Henderson, B.W., Jori, G., Kessel, D., Korbelik, M., Moan, J. and Peng, Q. (1998). Photodynamic therapy. *Journal of the National Cancer Institute, 90*, 889–905.

Drzewiecka-Matuszek, A., Skalna, A., Karocki, A., Stochel, G.and Fiedor, L. (2005). Effects of heavy central metal on the ground and excited states of chlorophyll. *Journal of Biological Inorganic Chemistry, 10*, 453-462.

Faller, P., Maly, T., Rutherford, A.W. and Macmillan, F. (2001). Chlorophyll and carotenoid radicals in photosystem II studied by pulsed ENDOR. *PubliCEA*. 320-326.

Ferro, S., Ricchelli, F., Mancini, G., Tognon, G. and Jori, G. (2006). Inactivation of methicillin-resistant Staphylococcus aureus (MRSA) by liposome-delivered photosensitising agents. *Journal of Photochemistry and Photobiology B: Biology, 83*, 98–104.

Fiedor, J., Fiedor, L., Kammhuber, N., Scherz, A. and Scheer, H. (2002). Photodynamics of the bacteriochlorophyll-carotenoid system - 2. influence of central metal, solvent and b-carotene on photobleaching of bacteriochlorophyll derivatives. *Photochemistry and Photobiology, 76*, 145-152.

Fiedor, L., Rosenbach-Belkin, V., Sai, M. and Scherz, A. (1996). Preparation of tetrapyrrole-amino acid covalent complexes. *Plant Physiology and Biochemistry, 34*, 393-398.

Fischer, B.B., Dayer, R., Wiesendanger, M. and Eggen, R.I.L. (2007). Independent regulation of the GPXH gene expression by primary and secondary effects of high light stress in *Chlamydomonas reinhardtii*. *Physiologia Plantarum, 130*, 195–206.

Foote, C.S. (1976). *Photosensitized oxidation and singlet oxygen: Consequences in biological systems*. In W.A. Pryor (Ed.) *Free radicals in biology II* (pp.85-133). New York, USA: Academic Press.

Foote, C.S. (1996). Definition of Type I and Type II photosensitized oxidation. *Photochemistry and Photobiology, 54*, 659-660.

Garrard, L.A., Van, T.K. and West, S.H. (1976). Plant response to middle ultraviolet (UV-B) radiation: Carbohydrate level and chloroplast reactions. *Proceedings - Soil and Crop Science Society of Florida, 36*, 184-188.

Girotti, A. (2001). Photosensitized oxidation of membrane lipids: reaction pathways, cytotoxic effects, and cytoprotective mechanisms. *Journal of Photochemistry and Photobiology B: Biology, 63*, 103–113.

Gouterman, M. (1978). Electronic Spectra. In D. Dolphyn (Ed.) *The Porphyrins: Physical chemistry-part A* (pp. 1-166). New York, USA: Academic Press.

Govindjee (2004). Chlorophyll a fluorescence: a bit of basics and history. In G.C. Papageorgiou, Govindjee (Eds.) *Advances in Photosynthesis and Respiration - Chlorophyll a Fluorescence* (pp. 1–42). Dordrecht, The Netherlands: Springer.

Green, B.R., Anderson, J.M. and Parson, W.W. (2003). Photosynthetic membranes and their light-harvesting antennas. In B.R. Green and W.W. Parson (Eds.) *Advances in photosynthesis and respiration: Light harvesting antennas in photosynthesis* (pp.1-28). Dordrecht, The Netherlands: Kluwer Academic Publishers.

Hakala, M., Tuominen, I., Kranen, M., Tyystjarvi, T. and Tyystjarvi, E. (2005). Evidence for the role of the oxygen-evolving manganese complex in photoinhibition of photosystem II. *Biochimica et Biophysica Acta, 1706*, 68-80.

Harbour, J.R. and Bolton, J.R. (1978). The involvement of the hydroxyl radical in the destructive photooxidation of chkorophylls *in vivo* and *in vitro*. *Photochemistry and Photobiology, 28*, 231-234.

Hendry, G.F., Houghton, J.F. and Brown, S.B. (1987). The degradation of chlorophyll – a biological enigma. *New Phytologist, 107*, 255-302.

Hideg, E., Kálai, T., Hideg, K. and Vass, I. (1998). Photoinhibition of photosynthesis in vivo results in singlet oxygen production detection via nitroxide-induced fluorescence quenching in broad bean leaves. *Biochemistry, 37*, 11405-11411.

Hideg, E., Takttsy, A., Stir, C.P., Vass, I. and Hideg, K. (1999). Utilizing new adamantyl spin traps in studying UV-B-induced oxidative damage of photosystem II. *Journal of Photochemistry and Photobiology B: Biology, 48*, 174-179.

Hirabayashi, H., Amakawa, M., Kamimura, Y., Shino, Y., Satoh, H., Itoh, S. and Tamiaki, H. (2006). Analysis of photooxidized pigments in water-soluble chlorophyll protein complex isolated from *Chenopodium albu*. *Journal of Photochemistry and Photobiology A: Chemistry 183*, 121-125.

Hoff, A.J. and Amesz, J. (1991). Visible absorption spectroscopy of chlorophylls. In H. Scheer (Ed.) *Chlorophylls* (pp. 723-738). Boca Raton, FL: CRC Press.

Holm-Hansen, O., Lubin, D., Helbling, E.W. (1993). Ultraviolet radiation and its effects on organisms in aquatic environments. In A.R. Young, L.O. Björn, J. Moan and W. Nultsch (Eds.) *Environmental UV photobiology* (pp. 379-426). New York, USA: Plenum Press.

Huang, C., Chen, J., Hu ,X., Liao, X., Yhang, Y. and Wu, J. (2008). Kinetics for photo-degradation of chlorophyll in postharvest fruit and vegetable. *Transactions of the Chinese Society of Agricultural Engineering, 24*, 233-238.

Hynninen, P.H. (1991). Chemistry of chlorophylls: Modifications. In H. Scheer (Ed.) *Chlorophylls* (pp. 145-209). Boca Raton, FL: CRC Press.

Inhoffen, H.H. (1968). Recent progress in chlorophyll and porphyrin chemistry. *Pure Applied Chemistry, 17*, 443-460.

Jansen, M.A.K., Gaba, V. and Greenberg, B.M. (1998). Higher plants and UV-B radiation: balancing damage, repair and acclimation. *Trends in Plant Science, 3*, 131-135.

Jeffrey, S.W., Mantoura, R.F.C. and Wright, S.W. (1996). Data for the identification of 47 key phytoplankton pigments. In *Phytoplankton pigments in oceanography: guidelines to modern method* (pp. 1-615). UNESCO Publishing.
Jen, J.J. and MacKinney, G. (1970): "On the decomposition of chlorophyll *in vitro* – II. Intermediates and breakdown products. *Photochemistry and Photobiology, 11*, 303-308.
Johnson, G.A. and Day, T.A. (2002). Enhancement of photosynthesis in Sorghum bicolor by ultraviolet radiation. *Physiologia Plantarum, 116*, 554-562.
Juzeniene, A., Peng, Q. and Moan, J. (2007). Milestones in the development of photodynamic therapy and fluorescence diagnosis. *Photochemical and Photobiological Sciences, 6*, 1234-1245.
Karukstis, K.K. (1991). Chlorophyll fluorescence as a physiological rpobe of the photosynthesis apparatus. In H. Scheer (Ed.) *Chlorophylls* (pp. 769-795). Boca Raton, FL: CRC Press.
Katz, J.J., Closs, G.L., Pennington, F.C., Thomas, M.R. and Strain H.H. (1963). Infrared spectra, molecular weights, and molecular association of chlorophylls a and b, methyl chlorophyllides, and pheophytins in various solvens. *Journal of the American Chemical Society, 85*, 3801-3809.
Katz, J.J., Shipman, L.L., Cotton, T.M. and Janson, T.R. (1978). Chlorophyll aggregation: coordination interactions in chlorophyll monomers, dimers and oligomer. In D. Dolphyn (Ed.), *The Porphyrins: Physical chemistry-part C* (pp. 401-458), New York, USA: Academic Press.
Kobayashi, M., Akiyama, M., Kano, H. and Kise, H. (2006). Spectroscopy and structure determination. In B. Grimm, R.J. Porra, W. Rüdier and H. Scheer (Eds.) *Advances in photosynthesis and respiration: Chlorophylls and Bacteriochlorophylls* (pp. 79–94). Dordrecht, The Netherlands: Springer.
Krasnovsky, A.A. Jr. (1994). Singlet molecular oxygen and primary mechanisms of photooxidative damage of chloroplasts: studies based on detection of oxygen and pigment phosphorescence. *Proceedings of the Royal Society of Edinburgh, Section B: Biological science, 102*, 219–235.
Larkum, A.W.D., Karge, M., Reifarth, F., Eckert, H-J., Post, A. and Renger, G. (2001). Effect of monochromatic UV-B radiation on electron transfer reactions of photosystem II. *Photosynthesis Research, 68*, 49-60.
Latouche, G., Cerovic, Z.G., Montagnini, F. and Moya, I. (2000). Light-induced changes of NADPH fluorescence in isolated chloroplasts: a

spectral and fluorescence lifetime study. *Biochimica and Biophysica Acta, 1460,* 311-329.
Lee, M. and Lee, W-Y. (1990). Anti retroviral effect of chlorophyll derivatives (CpD-D)by photosensitization. *Yonsei Medical Journal, 31,* 339-346.
Li, W-T., Tsao, H-W., Chen, Y-Y., Cheng, S-W. and Hsu, Y-C. (2007). A study on the photodynamic properties of chlorophyll derivatives using human hepatocellular carcinoma cells. *Photochemical and Photobiological Sciences, 6,* 1341–1348.
Lim, D-S., Ko, S-H., Kim, S-J., Park, Y-J., Park, J.-H. and Lee, W-Y. (2002). Photoinactivation of vesicular stomatitis virus by a photodynamic agent, chlorophyll derivatives from silkworm excreta. *Journal of Photochemistry and Photobiology B: Biology, 67,* 149–156.
Lindsay-Smith, J.R. and Calvin, M. (1966). Studies on chemical and photochemical oxidation of bacteriochlorophyll. *Journal of the American Chemical Society, 88,* 4500-4506.
Livingston, R. and Stockmann, D. (1962), A further study of the phototropy of chlorophyll in solution. *The Journal of Physical Chemistry B: Biology, 66,* 2533-2537.
Llewellyn, C.A., Fauzi, R., Mantouro, C. and Brereton, R.G. (1990a). Products of chlorophyll photodegradation - 1. Detection and separation. *Photochemistry and Photobiology, 52,* 1037-1041.
Llewellyn, C.A., Mantoura, R.F.G. and Brereton, R.G. (1990b). Products of chlorophyll photodegradation - 2. Structural identification. *Photochemistry and Photobiology, 52,* 1043-1047.
Markovic, D., Proll, S., Bubenzer, C. and Scheer, H. (2007). Myoglobin with chlorophyllous chromophores: influence on protein stability. *Biochimica et Biophysica Acta, 1767,* 897-904.
McDaniel, D.H. (2007). Method and apparatus for acne treatment. *US Patent 7,201,765 B2.*
Melis, A., Nemson, J.A. and Harrison, M.A.(1992). Damage to functional components and partial degradation of photosystem II reaction center proteins upon chloroplast exposure to ultraviolet-B radiation. *Biochimica et Biophysica Acta, 1100,* 312-320.
Merzlyak, M.N. and Hendry, G.A.F. (1994). Free radical metabolism, pigment degradation, and lipid peroxidation in leaves during senescence. *Proceedings-Royal Society of Edinburgh. Section B: Biological science, 102,* 459-471.
Merzlyak, M.N. (1989). Activated oxygen and oxidative processes in plant cell membranes. *Itogi Nauki i Tekhniki, Ser. Fiziologia Rastenija, Vol.6.*

Merzlyak, M.N., Kovrizhnykh, V.A. and Timofeev, K.N. (1991). Superoxide-mediated chlorophyll allomerization in a dimethyl sulfoxide-water mixture. *Free Radical Research Communications, 15*, 197-201.

Merzlyak, M.N., Pogosyan, S.I., Lekhimena, L., Zhigalova, T.V., Khozina, I.F., Cohe, Z. and Khrushchev, S.S. (1996). Spectral characterization of photooxidation products formed in Chl solution and upon photodamage to the cyanobacterium *Anabaena variabilis*. *Russian Journal of Plant Physiology, 43*, 160–168.

Morgan, J., Jackson, J.D., Zheng, X., Pandey, S.K. and Pandey, R.K. (2010). Substrate affinity of photosensitizers derived from chlorophyll-a: The ABCG2 transporter affects the phototoxic response of side population stem Cell-like cancer cells to photodynamic therapy. *Molecular Pharmaceutics, 7*, 1789-1804.

Morris, M.M., Park, Y. and MacKinney, G. (1973). On the decomposition if chlorophyll *in vitro*. *Journal of Agricultural and Food Chemistry, 21*, 277-279.

Moser, J.G. (1998). Definitions and general properties of 2nd snd 3rd generation photosensitizers, In J.G. Moser (Ed.) *Photodynamic tumor therapy. 2nd and 3rd generation photosensitizers* (pp.3-7). Amsterdam, The Netherlands: Harwood Academic Publishers.

Moser, J.G., Rueck, A., Westphal-Froesch, C. and Schwarzmayer, H.J. (1992). Photodynamic cancer therapy: fluorescence localization and light absorption spectra of chlorophyll-derived photosensitizers inside cancer cells. *Optical Enginnering, 31*, 1441-1446.

Neale, P.J., Cullen, J.J., Lesser, M.P. and Melis, A.(1993). Physiological bases for detecting and predicting photoinhibition of aquatic photosynthesis by PAR and UV radiation. In H. Yamamoto and C.M. Smith (Eds.) *Photosynthetic responses to the environment* (pp.61-77). Washington, DC: American Society of Plant Physiology.

Nedbal, L. and Koblizek, M. (2006). Chlorophyll fluorescence as a reporter on in vivo electron transport and regulation in plants. In B. Grimm, R.J. Porra, W. Rüdier and H. Scheer (Eds.), *Advances in photosynthesis and respiration: Chlorophylls and Bacteriochlorophylls* (pp.517-519). Dordrecht, The Netherlands: Springer.

Oba, T., Mimuro, M., Wang, Z.Y., Nozawa, T., Yoshida, S. and Watanabe, T. (1997). Spectral characteristics and colloidal properties of chlorophyll a' in aqueous methanol. *Journal of Physical Chemistry B, 101*, 3261-3268.

Olszowka, D., Maksymiec, W., Krupa, Z. and Krawzyka, S. (2003). Spectral analysis of pigment photobleaching in photosynthetic antenna complec

LHCIIb. *Journal of Photochemistry and Photobiology B: Biology, 70*, 21-30.

Parson, W.W. and Nagarajan V. (2003). Optical spectroscopy in photosynthetic antennas. In B.R. Green and W.W. Parson (Eds.) *Advances in photosynthesis and respiration: Light harvesting antennas in photosynthesis* (pp.83-127). Dordrecht, The Netherlands: Kluwer Academic Publishers.

Pascal, A.A., Liu, Z., Broess, K., van Oort, B., van Amerongen, H., Wang, C., Horton, P., Robert, B., Chang, W. and Ruban, A. (2005). Molecular basis of photoprotection and control of photosynthetic light-harvesting. *Nature, 436*, 134-137.

Proll, S., Wilhelm, B., Robert, B. and Scheer, H. (2006). Myoglobin with modified tetrapyrrole chromophores: binding specificity and photochemistry. *Biochimica et Biophysica Acta, 1757*, 750-763.

Redmond, R.W. and Gamlin, J.N. (1999). A compilation of singlet oxygen yields from biologically relevant molecules. *Photochemistry and Photobiology, 70*, 391-475.

Renger, G., Völker, M., Eckert, H.J., Fromme, R., Hohm-Veit, S. and Graber, P. (1989). On the mechanism of photosystem II deterioration by UV-B irradiation. *Photochemistry and Photobiology, 49*, 97-105.

Roeder, B. (1998). Photobiophysical parameters. In J.G. Moser (Ed.) *Photodynamic tumor therapy. 2nd and 3rd generation photosensitizers* (pp.9-13). Amsterdam, The Netherlands: Harwood Academic Publishers.

Rontani, J-F., Baillet, G. and Aubert, C. (1991). Production of acyclic isoprenoid compounds during the photodegradation of chlorophyll a in seawater. *Journal of Photochemistry and Photobiology A: Chemistry, 59*, 369–377.

Rontani, J-F., Beker, B., Raphel, D. and Baillet, G. (1995). Photodegradation of chlorophyll phytyl chain in dead phytoplanktonic cells. *Journal of Photochemistry and Photobiology A: Chemistry, 85*, 137–142.

Rosenbach-Belkin, V., Chen, L., Fiedor, L., Salomon, Y. and Scherz, A. (1998). Chlorophyll and bacteriophyll derivatives as photodynamic agents. In J.G. Moser (Ed.) *Photodynamic tumor therapy. 2nd and 3rd generation photosensitizers* (pp.117-125). Amsterdam, The Netherlands: Harwood Academic Publishers.

Rosenkrantz, A.A., Lunin, V.G., Gulak, P.V., Sergienko, O.V., Shumiantseva, M.A., Voronina, O.L., Gilyazova, D.G., John, A.P., Kofner, A.A., Mironov, A.F., Jans, D.A. and Sobolev, A.S. (2003). Recombinant modular transporters for cell-specific nuclear delivery of locally acting drugs enhance photosensitizer activity. *The FASEB Journal, 17*, 1121-1123.

Ruban, A.V., Berera, R., Ilioaia, C., van Stokkum, I.H.M., Kennis, J.T.M., Pascal, A.A., van Amerongen, H., Robert, B., Horton, P. and van Grondelle, R. (2007). Identification of a mechanism of photoprotective energy dissipation in higher plants. *Nature, 450*, 575-578.

Santabarbara, S. (2006). Limited sensitivity of pigment photo-oxidation in isolated thylakoids to singlet excited state quenching in photosystem II antenna. *Archives of Biochemistry and Biophysics, 455*, 77-88.

Sauer, K. and Debreczeny, M. (1996). Fluorescence. In J. Amesz and A.J. Hoff (Eds.) *Advances in Photosynthesis: Biophysical techniques in photosynthesis* (pp.41-61). Dordrecht, The Netherlands: Kluwer Academic Publishers.

Scheer, H. (1991). Structure and occurrence of chlorophylls. In H. Scheer (Ed.), *Chlorophylls* (pp. 1-30). Boca Raton, FL: CRC Press.

Scheer, H. (1994). Chemistry and spectroscopy of chlorophylls. In W. Horspool and S. Soon Song (Eds.), *CRC Handbook of organic photochemistry and photobiology* (pp. 1402-1411). Boca Raton, FL: CRC Press.

Scheer, H. (2003a). Chemistry and spectroscopy of chlorophylls. In W. Horspool and F. Lenci (Eds.), *CRC Handbook of organic photochemistry and photobiology* (pp. 1-16). Boca Raton, FL: CRC Press.

Scheer, H. (2003b). The Pigments. In B.R. Green and W.W. Parson (Eds.), *Advances in photosynthesis and respiration: Light harvesting antennas in photosynthesis* (pp. 29-81). Dordrecht, The Netherlands: Kluwer Academic Publishers.

Scheer, H. (2006). An overview of chlorophylls and bacteriochlorophylls: Biochemistry, biophysics, functions and applications. In B. Grimm, R.J. Porra, W. Rüdier and H. Scheer (Eds.), *Advances in photosynthesis and respiration: Chlorophylls and Bacteriochlorophylls* (pp. 1-26). Dordrecht, The Netherlands: Springer.

Scherz, A., Salomon, Y. and Fiedor, L. (1994). Chlorophyll and bacteriophyll derivatives, preparation and pharmaceutical compositions comprising them as photosensitizers for photodynamic therapy. *Chemical Abstracts, 120*, 386.

Schlichter, J., Friedrich, J., Parbel, M. and Scheer, H. (2000). New concepts in spectral diffusion physics of proteins. *Photonics Science News, 6*, 100-110.

Shaposhnikova, M.G. and Krasnovsky, A.A. (1973). Comparative investigation of the photooxidation of chlorophyll analogs in aqueous solutions of detergents. *Biokhimiya (Moscow), 38*, 193–200.
Shieh, M-J., Peng, C-L., Lou, P-J., Chiu, C-H, Tsai, T-Y., Hsu, C-Y., Yeh, C-Y and Lai, P-S. (2008). Non-toxic phototriggered gene transfection by PAMAM-porphyrin conjugates. *Journal of Controlled Release, 129*, 200–206.
Simonich, M.T., Egner, P.A., Roebuck, B.D., Orner, G.A., Jubert, C., Perreira, C., Groopman, J.D., Kensler, T.W., Dashwood, R.H., Williams, D.E. and Baily, G.S. (2007). Natural chlorophyll inhibits aflatoxin B1-induced multi-organ carcinogenesis in the rat. *Carcinogenesis 28*, 1294-1302.
Simonich, M.T., McQuistan, T., Jubert, C., Pereira, C., Hendricks, J.D., Schimerlik, M., Zhu, B., Dashwood, R.H., Williams, D.E. and Baily, G.S. (2008). Low-dose dietary chlorophyll inhibits multi-organ carcinogenesis in the rainbow trout. *Food and Chemical Toxicology, 46*, 1014–1024.
Steiner, R., Cmiel, E. and Scheer, H. (1983). Chemistry of bacteriochlorophyll b: identification of some (photo)oxidation products. *Zeitschrift fur Naturforschung C, 38C*, 748-752.
Sternberg, E.D. and Dolphin, D. (1998). Porphyrin-based photosensitizers for use in photodynamic therapy. *Tetrahedron, 54*, 4151-4202.
Strid, A. and Porra, R.J. (1992). Alteration in pigment content in leaves of *Pisum sativum* after exposure to supplementary UV-B. *Plant Cell Physiology, 33*, 1015-1023.
Struck, A., Cmiel, E., Schneider, S. and Scheer, H. (1990). Photochemical ring-opening in meso-chlorinated chlorophylls. *Photochemistry and Photobiology, 52*, 217-222.
Teramura, A. and Ziska, L. (1996). Ultraviolet-B radiation and photosynthesis. In N. Baker (Ed.) *Advances in photosynthesis: Photosynthesis and the environment* (pp. 435-450). Dordrecht, The Netherlands: Kluwer Academic Publishers.
Tevini, M., Iwanzik, M.W. and Thoma, U. (1981). Some effects of enhanced UV-B irradiation on the growth and composition of plants. *Planta, 153*, 388-394.
Trifunac, A.D. and Katz, J.J. (1974). Structure of chlorophyll a dimers in solution from proton magnetic resonance and visible absorption spectroscopy. *Journal of the American Chemical Society, 96*, 5233-5240.
Troxler, R.F., Smith, K.M. and Brown, S.B. (1980). Mechanism of photooxygenation of bacteriophyll-c derivatives. *Tetrahedron Letters, 21*, 491-494.

Turcsányi, E. and Vass, I. (2000). Inhibition of photosynthetic electron transport by UV-A radiation targets the photosystem II complex. *Photochemistry and Photobiology, 72*, 513–520.

van Amerongen, H. and Dekker J.P. (2003). Light-harvesting in photosystem II. In B.R. Green and W.W. Parson (Eds.) *Advances in photosynthesis and respiration: Light harvesting antennas in photosynthesis* (pp.219-251). Dordrecht, The Netherlands: Kluwer Academic Publishers.

Vass, I. (1997). Adverse effects of UV-B light on the structure and function of the photosynthetic apparatus. In M. Pessarakli (Ed.) *Handbook of photosynthesis* (pp. 931–949). New York, USA: Marcel Dekker Inc.

Vass, I., Turcsányi, E., Touloupakis, E., Ghanotakis, D. and Petroluleas, V. (2002). The mechanism of UV-A radiation-induced inhibition of photosystem II electron transport studied by EPR and chlorophyll fluorescence. *Biochemistry, 41*, 10200-10208.

Vincent, W.F. and Neale, P.J. (2000): Mechanisms of UV damage to aquatic organisms. In S.I.de Mora, S. Demers and M.Vernet, (Eds.) *The effects of UV radiation in the marine environment* (pp.149-176). Cambridge, UK: Cambridge University Press.

Vu, C.V., Allen, L.H.Jr. and Garrard, L.A. (1984). Effects of UV–B radiation (280–320 nm) on ribulose 1,5-bisphosphate carboxylase in pea and soybean. *Environmental and Experimental Botany, 24*, 131-143.

Walker, D.A., Cerović, Z.G. and Robinson, S.P. (1987). Isolaton of intact chloroplasts: general principles and criteria of integrity. *Methods in Enzimology, 148*, 145-157.

White, R.A. and Tollin, G. (1971). Chlorophyll one-electron photochemistry: light-induced absorbance changes and ESR signals for various porpharin-quinone and hydroquinone systems in alcohol solvents. *Photochemistry and Photobiology, 14*, 15-42.

Whitmarsh, J. and Govindjee (1995). Photosynthesis. In G.L. Trigg and E.H. Immergut (Eds.) *Encyclopedia of applied physics* (pp. 513-532). New York, USA: VCH Publishers, Inc.

Wilson, D.F. and Cerniglia, G.J. (1994). Oxygenation of tumors as evaluated by phosphorescence imaging. *Advances in Experimental Medicine and Biology, 345*, 539-547.

Wondrak, G.T., Jacobson, M.K. and Jacobson, E.L. (2006). Endogenous UVA-photosensitizers: mediators of skin photodamage and novel targets for skin photoprotection. *Photochemical and Photobiological Sciences, 5*, 215–237.

Woodward, R.B. (1961). The total synthesis of chlorophyll. *Pure Applied Chemistry, 2*, 383-404.

Yavlovich, A., Singh, A., Blumenthal, R. and Puri, A. (2011). A novel class of photo-triggerable liposomes containing DPPC:DC8,9PC as vehicles for delivery of doxorubcin to cells. *Biochimica et Biophysica Acta, 1808*, 117–126.

Yujing, M. and Mellouki, A. (2000). The near-UV absorption cross sections for several ketones. *Journal of Photochemistry and Photobiology A: Chemistry, 134*, 31-36.

Zhang, Z., Tashiro, Y., Matsukawa, S. and Ogawa, H. (2005). Influence of pH and temperature on the ultraviolet-absorbing properties of porphyra-334. *Fisheries Science, 71*, 1382-1384.

Zvezdanović, J. and Marković, D. (2008). Bleaching of chlorophylls by UV irradiation *in vitro*: the effects on chlorophyll organization in acetone and *n*-hexane. *Journal of the Serbian Chemical Society, 73*, 271-282.

Zvezdanović, J., Cvetić, T., Veljović-Jovanović, S. and Marković, D. (2009). Chlorophyll bleaching by UV-irradiation *in vitro* and *in situ*: Absorption and fluorescence studies. *Radiation Physics and Chemistry, 78*, 25-32.

In: Chlorophyll
Editors: H. Le, et.al.

ISBN: 978-1-61470-974-9
© 2012 Nova Science Publishers, Inc.

Chapter II

From Protochlorophillide to Chlorophyll: Final Light-Dependent Stage in Biosynthesis of Main Photosynthetic Pigment

Olga B. Belyaeva[*] *and Felix F. Litvin*
Faculty of Biology, Lomonosov Moscow State University, Moscow, Russia

Abstract

The key step in chlorophyll biosynthesis is photoreduction of its immediate precursor, protochlorophyllide. This reaction consists in the attachment of two hydrogen atoms in positions C17 and C18 of the tetrapyrrole molecule of protochlorophyllide; the double bond is replaced with the single bond.

The purpose of this survey is to summarize and discuss the data obtained in the studies on organization of the active pigment–protein complex where the photoreduction of protochlorophyllide proceeds, the mechanisms of the primary photophysical and photochemical reactions of

---

[*] Corresponding author: Fax: +7 (495) 939-54-89; E-Mail: olgabelyaeva@mail.ru.

this process and the pathways of functionally chlorophyll species formation.

The active pigment–protein complex includes protochlorophyllide, the donor of hydride ion NADPH, and the photoenzyme protochlorophyllide oxidoreductase (POR). In the living cell there are several spectrally different active forms of the chlorophyllide precursor. Based on the results of numerous investigations, it can be stated that the reduction of the active forms of protochlorophyllide is a multistep process comprising two or three short-living intermediates characterized by the singlet ESR signal. The first intermediate seems to be a complex with a charge transfer between protochlorophyllide and the donor of hydride ion (NADPH). The donor of the second proton is the tyrosine residue Tyr 193 of the photoenzyme. The photoreduction of protochlorophyllide is preceded by light-stimulated conformational changes in the enzyme active site, enabling the hydride and proton transfer reactions to occur.

Formed as a result of the photoreaction, the primary forms of chlorophyllide undergo further dark and light-dependent transformations, sequential and parallel, leading to the formation of different forms of chlorophyll and pheophytin that start the formation of the pigment apparatus of the two photosystems and light-harvesting antenna.

Introduction

Chlorophyll biosynthesis in plant leaves is one of the most important biospheric processes. It enables production, renewal, and permanent maintenance of the main photosynthetic pigment required for effective photosynthesis and high productivity of plants. Annual synthesis of chlorophyll in the biosphere amounts to more than one billion metric tons.

In the absence of light chlorophyll biosynthesis ceases at the stage of protochlorophyllide accumulation in etiolated leaves. This precursor (see Figure 1) differs from chlorophyll by the absence of a phytol moiety and by the presence of a double bond in the macrocyclic D-ring. The terminal light-dependent stage of chlorophyll biosynthesis begins from the fast and highly efficient photochemical reduction of the double bond, which initiates the network of dark and photochemical reactions leading to the production of several chlorophyll–protein complexes and to their incorporation into the structures that participate in the construction of photosynthetic pigment machinery.

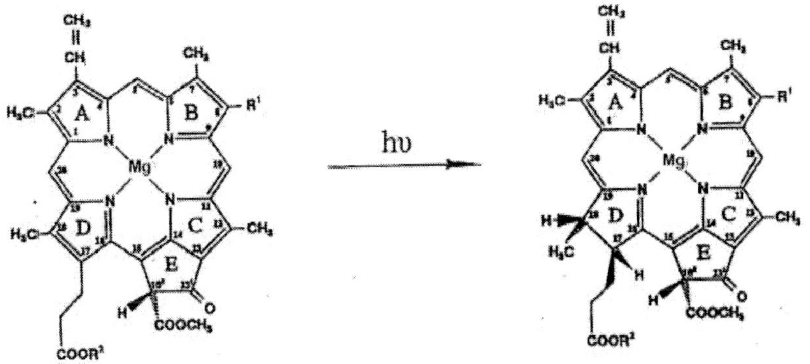

Figure 1. Structures of protochlorophyllide and chlorophyll(ide) molecules. R1 designates the CH2-CH3 group for the monovinyl form of the pigment and symbolizes CH=CH2 for the divinyl form. R2 designates C20H39 (phytol) in the case of protochlorophyll or chlorophyll and stands for hydrogen atom in the case of protochlorophyllide and chlorophyllide.

A characteristic feature of chlorophyll formation is its exclusive occurrence in integral biological systems such as leaves and leaf homogenates. The photochemical conversion of protochlorophyll(ide) to chlorophyll(ide) in simple systems (solutions) has not been realized so far. Already the earliest studies established that chlorophyll biosynthesis occurs in specialized native complexes of the pigment and some carrier providing for physiological activity and spectral properties *in vivo*.

Photoactive Pigment–Enzyme Complexes of the Chlorophyll Precursor

Over the past few decades, the advance in deciphering the main mechanism for the terminal stage of chlorophyll biosynthesis was related with identification of the hydrogen donor (NADPH) in the reaction of protochlorophyllide photoreduction [1, 2] and with the discovery of protochlorophyllide oxidoreductase (POR), an enzyme catalyzing the photoreaction of protochlorophyllide conversion to chlorophyllide [1-6]. The conclusion was drawn that the photoreduction of protochlorophyllide proceeds in a specialized active complex comprising, apart from the enzyme, the reducible protochlorophyllide and the hydrogen donor NADPH.

Photoenzyme Protochlorophyllide Oxidoreductase

The analyses of the POR amino acid sequence [7, 8] and the POR secondary structure [9] led to the conclusion that this enzyme belongs to the family of short-chain alcohol dehydrogenases within the enzyme superfamily "RED" (reductases – epimerases – dehydrogenases). Two features distinguish POR from other members of short-chain dehydrogenase–reductase family: porphyrin is used as a substrate, and the enzyme activity is strictly light dependent. It is presently established that POR is a unique photoenzyme occurring in all plant species, from cyanobacteria to higher plants [10-12]. POR is encoded in the nucleus and translated to the cytoplasm as a high-molecular-weight precursor (41–44 kD) [13], which is then imported into the plastids through the envelope membranes [13, 14]. The transit peptide (about 8 kD) is cleaved off by the stromal peptidase, giving rise to the mature enzyme with the molecular mass of about 36 kD.

The unique feature of POR compared to other members of the RED superfamily is the existence of an external hydrophobic loop (33 amino acid residues) between the fifth and the sixth β-strands. The exact function of a broad hydrophobic loop is unclear. However, it was proposed that this loop is significant for protochlorophyllide binding, for connection with the membrane (anchoring of the enzyme), and for the formation of the enzyme dimers [10]. A three-dimensional structural model for the ternary complex Pchlide–NADPH–POR was suggested based on structural homology between POR and the RED superfamily proteins [15].

The characteristics of the suggested model correspond to the globular water-soluble protein, whose structure contains (in the N-terminal domain) a typical narrow cleft that binds the cofactor NADPH ("Rossman fold"). The native photenzyme POR occurs predominantly in the dimeric form [16, 17]. It represents a peripheral membrane protein localized on the stromal side of the thylakoid membrane [10, 14, 18-24].

POR Active Site

In the family of short-chain alcohol dehydrogenases, the substrate is reduced by the hydride ion supplied from NADPH that is bound at the bottom of the substrate-binding cleft near conservative tyrosine and lysine residues involved in the proton transfer to the substrate. Since the photoenzyme POR is structurally similar to short-chain alcohol dehydrogenases, it seemed likely that the reaction mechanism of protochlorophyllide photoreduction involves the transfer of hydride ion from NADPH, like it occurs during the substrate reduction by alcohol dehydrogenases. When the protochlorophyllide photoreduction in the reaction mixture containing protochlorophyllide, NADPH, and the etioplast membranes was examined by means of NMR spectroscopy using radioactive hydrogen (^{3}H), the results showed that the hydride ion arriving from the 4S position of NADPH nicotinamide moiety attaches to the C_{17} position of the protochlorophyllide molecule, thus accomplishing the trans-reduction of the pigment molecule [25, 26]. The notion that the hydride ion and the proton transfer occur asynchronously during photoreduction of protochlorophyllide molecule is evidenced from observations of several short-lived intermediates (see below). The second proton is apparently donated from the conserved residue Tyr193 located in the vicinity of C17=C18 bond [8, 27, 28]. According to the model proposed by Wilks and Timko [8] (Figure 2), the position of the protochlorophyllide ring D is fixed against NADPH and the tyrosine residue, which enables the hydride ion and proton transfer. The optimal structure of the ternary complex satisfying the steric requirements for the reaction of protochlorophyllide photoreduction depends largely on the conserved residues Lys197 and Cys226 [8, 27, 28]. The proton of the tyrosine phenol group is transferred to the C_{18} atom of the protochlorophyllide molecule. In reconstituted ternary complexes comprising the mutant POR, whose Tyr-193 was replaced with Phe, the protochlorophyllide photoreduction ceased at the step of intermediary photoproduct, and the second (light-independent) stage of the reaction was inactivated [27]. These data demonstrate that hydride ion is transferred prior to the proton transfer.

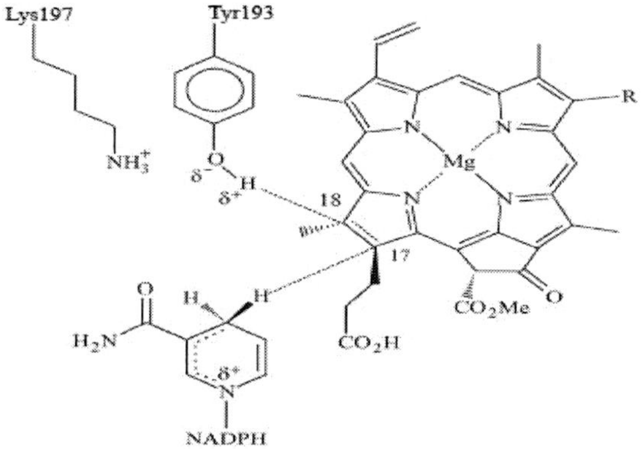

Figure 2. Suggested model of the catalytic mechanism for protochlorophyllide oxidoreductase in the active center of ternary complex Pchlide–POR–NADPH. Interaction of complex components is shown as in [8].

The binding of protochlorophyllide molecules to the enzyme POR active center and the photochemical activity of the complex depend crucially on three sites of the pigment molecule, namely, the central Mg^{2+} atom [4, 29], the side chain of the ring D, and the ring E structure [24, 29, 30].

It was found that isolated POR is capable of reducing protochlorophyllide *a* but not protochlorophyll *a* (esterified form) [1, 4, 31, 32]. The side group at C17 position should be propionic acid holding a free carboxyl group. Longer alcohol chains like phytol or an acrylic group prevent the photoreduction of the pigment [24, 33]. However, some exceptions from this rule were observed. In the study [34], the POR isolated from C-2A mutant of green alga *Scenedesmus obliquus* was competent in reducing protochlorophyllide esterified by long-chain alcohol. According to the report [35], the long-wavelength esterified protochlorophyll form (Pchl 682/672) in *Chlorella* cells was capable of conversion to chlorophyll. The authors proposed that algae probably contain a specific POR form that catalyzes photoreduction of esterified molecule of chlorophyll precursor. It should be noted that the ability of esterified protochlorophyllide to produce chlorophyll was also observed in earlier studies [36-38], but the efficiency of this transformation was very low. Therefore, one may suppose the existence of a minor pool of active complexes including the chlorophyll precursor esterified with phytol. Interestingly, the esterified protochlorophyllide is mainly located in prothylakoids [39].

Protochlorophyllide with chemical modifications of the ring E in positions $C13^1$ and $C13^2$ is unable to be a substrate for POR [24]. The changes in stereochemistry of -H and -CO_2CH_3 groups in the $C13^2$ position [24, 40] also led to inhibition of protochlorophyllide photoreduction reaction. Recent studies concerning the spectral changes of reconstituted ternary complexes in the infrared region have shown that protochlorophyllide is linked to POR by means of strong H-bonding interaction between the keto group in $C13^1$ position and protein residues or a water molecule and two coordination interactions between protein residues or water molecule/s and the Mg atom of protochlorophyllide [29].

Heterogeneity of Protochlorophyllide Oxidoreductase

Plant leaves contain three forms of protochlorophyllide oxidoreductase (PORA, PORB, and PORC) encoded by different genes (*PorA, PorB, PorC*) [41–48].

These proteins are quite similar in structure, as evidenced by high homology of their amino acid sequences [46] and by similar molecular masses (36, 37, and 38 kD, respectively).

At the same time, the synthesis of three POR species, like synthesis of corresponding mRNA, is differentially regulated by light. PORA is synthesized in darkness and constitutes the major part of paracrystalline prolamellar bodies in etioplasts. The transcription of *PorA* is strongly inhibited by light, and the enzyme POR is rapidly destructed by the light-induced protease within the first few hours of greening [49-51]. The transcription of the *PorB* gene proceeds both in darkness and in the light, with the ongoing subsequent translation into the enzyme PORB responsible for chlorophyll biosynthesis in the daylight. Transcription of *PorB* is stimulated by light during de-etiolation but is insensitive to irradiance upon growing plants under continuous white light [41, 42]. The transcription of *PorC* was indiscernible in darkness, but it increased during illumination and was stimulated by high-intensity light [46, 47, 52].

Multiplicity of Chlorophyll Precursor Forms in *Vivo*

The leaves of etiolated and green plants contain several spectrally different forms of protochlorophyllide that were identified using low-temperature and derivative spectroscopy and by means of spectra decomposition into constituent Gaussian components [53-56]. The forms of protochlorophyllide are usually marked with its maxima of fluorescence and absorbsion. The main spectral protochlorophyllide forms, i.e., Pchld633/628, Pchld643/637, and the routinely dominant form Pchld655/650, differ in terms of their photochemical activity. In addition, five minor long-wavelength forms were identified: Pchld666-669/658, Pchld680-682/668, Pchld690-692/677, Pchld698/686, and Pchld728/696.

The main photoactive form is protochlorophyllide Pchld655/650, which transforms into chlorophyllide even at very low temperatures. Two pools of this form were found out: Pchld653/648 and Pchld657/650 [55, 57]. The highest accumulation of photochemically active form Pchld653/648 was observed in very young (2–3 days) etiolated leaves, where the prolamellar bodies had not yet developed [58, 59].

Protochlorophyllide Pchld643/635 shows up in absorption spectra but is practically invisible in fluorescence spectra, which is caused by highly effective energy transfer from this form to Pchld655/650 [60-62]. The form Pchld643/637 is evident in young etiolated leaves [55] and is a dominant form in etiolated leaves of some species [63]. This form is the dominant one in homogenates of etiolated leaves and in reconstituted ternary complexes comprising protochlorophyllide, POR, and NADPH.

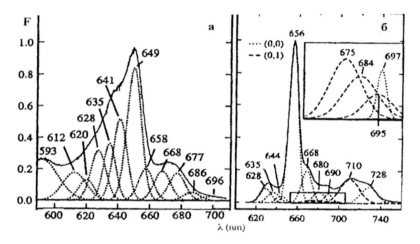

Figure 3. Decomposition of low-temperature fluorescence spectra (b) and fluorescence excitation spectra (a) of etiolated maize leaves into Gaussian components. Dotted lines (.....) designate electronic transitions; dashed lines (-----) designate vibrational components. Spectra of fluorescence excitation were measured for the emission at 740 nm [56]. (Reproduced by permission of the Royal Society of Chemistry (RSC) for the European Society for Photobiology, the European Photochemistry Association and the RSC).

The form Pchld643/637 is photochemically active and, similarly to protochlorophyllide Pchld655/650, transforms to chlorophyllide at rather low temperatures. However, its photochemical activity *in vivo* shows a stronger temperature dependence compared to photoactivity of Pchld655/650 [64]. The long-wavelength position of maxima for active protochlorophyllide forms Pchld643/637 and Pchld655/650 is likely caused not only by the linkage of the chromophore with the photoenzyme POR and the hydrogen donor NADPH but also by the chromophore–chromophore interactions between pigment molecules. Evidence of such interactions was obtained from studies of circular dichroism [65], fluorescence polarization, and energy transfer [66]. It is thought that protochlorophyllide molecules produce dimers [65] or tetramers [66].

The form Pchld633/628 is thought to be inactive because it does not transform to chlorophyllide upon short-term illumination [54, 67]. However, in plants enriched with this form, its slow conversion to chlorophyllide was observed at temperatures above 5°C [64, 68, 69]. Apparently, the monomeric short-wavelength protochlorophyllide Pchld633/628 is bound to the enzyme [70] but is not bound to NADPH [71]. According to the results of some studies, this protochlorophyllide form is the precursor of photoactive forms, Pchlide655/650 and P643/637 in the pathway of pigment dark synthesis [71, 72]. The active short-wavelength protochlorophyllide form occurs in juvenile (embryonic) leaves of dicotyledonous plants [59, 73, 74].

The native long-wavelength forms are supposedly related to the existence of large protochlorophyllide aggregates. This is evident from the occurrence of similar spectral bands in model systems with aggregated protochlorophyll [75-77] and in seed coats of some plants species [78-79] where the long-wavelength protochlorophyll species is present in the crystalline form [78]. The formation of protochlorophyllide aggregates is presumably facilitated by aggregation of the enzyme POR, which promotes the interactions between the porphyrin rings of the pigment.

The minor long-wavelength protochlorophyll(ide) forms, in addition to the main active forms, are involved in biosynthesis of chlorophyll functional species (see below).

Primary Photochemical and Photophysical Reactions in Photoreduction of Protochlorophyllide

Primary Reactions of Protochlorophyllide Photoreduction in Plant Leaves and Isolated Pigment–Protein Complexes

Labile Intermediates Stabilized at Low Temperatures

Analysis of protochlorophyllide photoreduction *in vivo* at very low temperatures, at which biochemical temperature-dependent steps are inhibited, helped to clarify the mechanism of this reaction. The quantum yield of protochlorophyllide fluorescence was found to decrease after illumination of etiolated leaves at low temperature (77–173 K), while the subsequent increase in the sample temperature led to the appearance of fluorescence bands characteristic of chlorophyll [80-90].

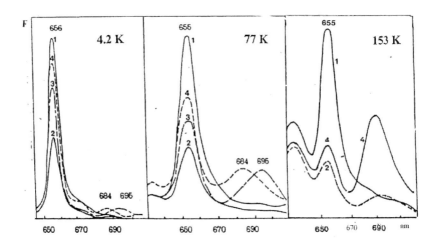

Figure 4. Fluorescence quenching as a primary act in photoreduction of the active protochlorophyllide species. Changes in low-temperature (77 K) fluorescence spectra of etiolated bean leaves after illumination at irradiance of 10^4 W/m^2 at low temperatures (4.2 K, 77 K, 153 K): *1* — nonirradiated etiolated leaf; *2* — the same leaf after irradiation at low temperature (temperature is indicated in the upper graph part); *3* — the same sample after warming to 233 K in darkness; *4* — the same sample after heating to 263 K in darkness.

The researchers supposed that the absorption of fluorescence-exciting light at low temperatures converts the protochlorophyllide molecule into the intermediary state (nonfluorescent intermediate), which is transformed to chlorophyllide during subsequent dark reaction proceeding after the temperature increase. The absorption band of nonfluorescent intermediate stabilized at low temperature is positioned near 690 nm [91]. The discovered intermediate was termed *X690*.

$$\text{Pchld655/650} \xrightarrow{h\nu} \textbf{X690} \rightarrow \text{Chld}$$

After comparing spectral changes induced by illumination of etiolated leaves at various temperatures (from 4.2 K to 153 K) [87, 92-94], the authors proposed that the intermediate *X690* formation is preceded by even earlier photochemical reactions.

The application of fluorescence and absorption spectroscopy revealed that the primary reaction of protochlorophyllide conversion into the nonfluorescent intermediate can be reversed under the action of powerful monochromatic laser beam (694 nm) [88]. The rate constants and the quantum yields were calculated both for the forward and backward reactions under the photostationary equilibrium. The calculations have proven that the low quantum yield of the whole process cannot be explained by photoreversibility of the reaction producing *X690*, because the quantum yield of the backward process was considerably (20 times) lower than that for the forward reaction. Since *X690* is stable in darkness, the authors supposed that there exits an additional intermediary step prior to *X690* formation and that this step comprises a fast backward reaction, which reduces the overall yield of the process. The putative intermediate was designated by symbol R (reversible):

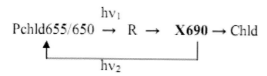

These results clarified the reasons for the smaller quantum yield of the reaction at low temperatures than at room temperature, at which it was about 0.5. Analysis of spectral changes led to the suggestion that the absorption spectrum of the primary intermediate R, whose production dominates at 77 K, is close or even identical to the absorption spectrum of the protochlorophyllide active form: the primary action of light on etiolated leaves, evident in quenching of protochlorophyllide fluorescence, occurs under conditions when *X690* is hardly discernible in absorption spectra [93].

Further evidence that the chain of protochlorophyllide transformations includes the short-lived intermediate with the absorption band similar to that of protochlorophyllide active form emerged from the time-resolved spectroscopy studies performed at room temperature [95, 96].

The early stages of protochlorophyllide photoreduction were also studied by comparing low-temperature (77 K) optical spectra (absorption and fluorescence) and ESR spectra of whole etiolated and illuminated leaves. These studies have shown that the two short-lived intermediates, *R* and *X690* exhibit paramagnetic properties [93]. After illumination of leaves at 77 K, when the production of nonfluorescent intermediate *R* was predominant, a structureless singlet ESR signal was observed with a bandwidth of 1.1. mT and a g-factor of 2.0021 characteristic of the free electron. As the sample temperature was raised gradually to 200 K, the ESR signal amplitude increased in parallel with narrowing of the signal width. Such a pattern of ESR signal changes apparently corresponds to the transformation of the primary intermediate *R* into the intermediate *X690*. Upon further increase in temperature above 200 K, the ESR signal amplitude dropped abruptly, reaching its initial dark level at the temperature of about 250 K. These changes occurred synchronously with the appearance of fluorescence spectral bands assigned to primary forms of chlorophyllide.

Paramagnetism of electron donors or acceptors, manifested in the light-induced ESR signal is one of the most convincing evidence that the charge-transfer complex is formed in the donor–acceptor system. A new absorption band emerges during formation of charge-transfer complex if oxidation–reduction potentials of interacting partners are adequate for complete electron transfer from one to the other molecule within the complex. When the complex with partial charge transfer is formed, the fluorescence quenching is observed without the appearance of new bands in the absorption spectrum. Judging from spectral characteristics of two nonfluorescing intermediates *in vivo* (R and *X690*), one may suppose that the photoinduced formation of these intermediates corresponds to the formation of complexes with partial charge transfer [$D^{\delta+} A^{\delta-}$] and with complete charge transfer [$D^+ A^-$].

The attempts to perform photoreduction of protochlorophyllide in solutions led to detection at 77 K of primary nonfluorescent products of protochlorophyll(ide) photoreduction whose spectral properties were similar to those of intermediates *R* and *X690* [93]. In diluted solutions the photoreaction product with spectral properties similar to those of intermediate *R* was stable only at low temperature (77 K). This intermediate transforms into the initial protochlorophyllide when the temperature increases above 200 K; i.e., the photoreaction is fully reversible. Irreversible photoreduction with the formation of *X690*-like product can proceed in concentrated solutions containing the aggregated pigment; however, the outcome of the reaction is low in this case, whereas the quantum yield of photoreduction in the native system (in whole etiolated leaves) is very high.

Primary Fluorescent Forms of Chlorophyllide

The nonfluoresecent intermediate *X690* produced under the action of light at low temperatures is converted to chlorophyllide in the dark reaction after the increase in temperature of the illuminated sample. When etiolated leaves were illuminated with white light at 77 K, the return to higher temperature was followed by almost simultaneous formation of two primary chlorophyllide species with fluorescence maxima at 684 and 695 nm and the respective absorption bands at 676 and 684 nm [87]. Thereafter, Chlide695/684 was converted in the dark reaction to Chlide684/676. The proportion of intensities for two bands arising after illumination at low temperature depended on spectral quality of actinic light. When blue light (with the peak intensity at 470 nm) was used for illumination, the increase in temperature resulted in the only fluorescence band at 695 nm, which shifted gradually to 684 nm. When the leaves were illuminated with red light (wavelengths above 600 nm), the temperature rise was followed by the appearance in the spectrum of a single short-wavelength maximum at 684 nm. In etioplast preparations only one short-wavelength primary chlorophyllide species was produced. These data implied that etiolated leaves contain two kinds of active protochlorophyllide–protein complexes exhibiting nearly identical absorption and fluorescence bands in the red spectral region. Both forms are converted to chlorophyllide through the stage of nonfluorescent intermediate formation. One of these precursors is able to transform immediately into the short-wavelength chlorophyllide form Chld684/676. The hypothetical reaction scheme was presented as follows:

$$\text{Pchld655/650} \xrightarrow{h\nu} R \rightarrow X690 \rightarrow \text{Chld695/684}$$
$$\text{Pchld655/650} \xrightarrow{h\nu} R \rightarrow X690 \rightarrow \text{Chld684/676} \rightarrow$$

The data based on time-resolved room temperature fluorescence spectroscopy [97], as well as on low-temperature fluorescence spectroscopy and deconvolution of spectra into the Gaussian components [98] have shown that the dark transformation of nonfluorescent intermediates gives rise to four primary chlorophyllide components with maxima at 684, 690, 695–697, and 706 nm. The results obtained by differential spectroscopy revealed that the long-wavelength primary chlorophyllide forms with fluorescence maxima at 696 and 706 nm are subsequently transformed at room temperature to the short-wavelength forms with fluorescence peaks at 675 and 684 nm, respectively. Apparently, the production of several primary labile chlorophyllide species indicates early differentiation of the pathways of functionally different native pigment species formation mediated by different protochlorophyllide forms.

Investigations of Fast Stages of Protochlorophyllide Photoreduction in Vivo at Physiological Temperatures

The short-lived intermediates of protochlorophyllide photoreduction can be detected *in vivo* by means of time-resolved spectroscopy at physiological temperatures.

The studies on the kinetics of fluorescence intensity [99] and fast changes in absorption spectra of etiolated leaves and isolated pigment–protein complexes [99, 100] at room temperature revealed the primary short-lived (0.2 µs) nonfluorescent intermediate with the absorption maximum at 690–695 nm.

By applying the nanosecond and picosecond absorption spectroscopy to analysis of protochlorophyllide photoreduction in isolated active pigment–protein complexes from etiolated leaves, Iwai and co-authors [95] detected at physiological temperature the formation of four intermediates with time constants of 50 ps, 2 ns, 35–250 ns, and 1–2 µs and the subsequent appearance of chlorophyllide (12 µs):

$$\text{Pchlde }(S_0) \xrightarrow{hv} \text{Pchlde }(S_1^*) \xrightarrow{\leq 50\text{ps}} \text{Pchlde }(S_1) \xrightarrow{1-2\text{ ns}} x_1 \xrightarrow{35-250\text{ ns}} x_2 \xrightarrow{1-2\text{ µs}} x_3 \xrightarrow{12\text{ µs}} \text{Chlde}$$
$$(=X_0)$$

Investigation of changes in differential absorption spectra has shown that the positions of absorption maxima for the intermediates X_0 and X_1 are nearly identical to the absorption band position of the initial protochlorophyllide ($\geq$ 640 nm), whereas the intermediates X_2 and X_3 are characterized by the absorption maxima at 688 and 684 nm, respectively. The time constant of X_0 formation corresponded to the relaxation time of the protochlorophyll molecule from the Franck–Condon state (S_1^*) to the equilibrium state (S_1); hence the X_0 entity is not an intermediate in the chemical sense. The intermediate X_1 is likely identical to the intermediate R discovered in the work [88], as evidenced from similarity of its absorption spectrum to the absorption spectrum of the initial protochlorophyllide. The intermediate X_2 is likely identical to the intermediate *X690*. This view is based on the absorption spectrum of this intermediate (maximum at 688 nm) and on its rise time (35–250 ns), which is comparable to the formation time of intermediate *X690* determined by other researchers [100, 101]. The intermediate X_3 can be compared with one of the primary forms of chlorophyllide arising in etiolated leaves preilluminated at 77 K and returned to higher temperature.

Thus, the comparative analysis of the research data concerning intermediate stages of protochlorophyllide photoreduction *in vivo* at physiological and low temperatures leads to the conclusion that this process comprises several intermediary products, including two or three short-lived intermediates characterized by strong quenching of protochlorophyllide fluorescence.

Elementary Reactions of Protochlorophyllide Photoreduction in Reconstituted Ternary Complexes

In the last few years a growing number of researchers began to study protochlorophyllide photoreduction by using artificial ternary complexes composed of protochlorophyllide, NADPH, and the photoenzyme protochlorophyllide oxidoreductase. The main active protochlorophyllide species in such systems is Pchlide644/642 [89, 90, 102, 103], which is similar in spectral properties to the protochlorophyllide active form Pchlide643/639 in whole etiolated leaves and to the main pigment species retained in isolated pigment–protein complexes.

Similarly to observations with whole etiolated leaves, the protochlorophyllide photoreduction in artificial ternary complexes at low temperatures was accompanied by the appearance of unstable nonfluorescent intermediate with the absorption band at 696 nm [19, 48], arising as a primary product after illumination of samples at low temperatures [89, 90, 103, 104]. The fluorescence quenching was paralleled by the appearance in the ESR spectrum of a singlet signal with the g-factor characteristic of the free electron [102, 104]. Upon the increase in temperature, the nonfluorescent intermediate was spontaneously converted into chlorophyllide, which was identified from its spectral properties and from the liquid chromatography analysis [102, 104]. The primary fluorescent intermediate was transformed in the dark reaction into the short-wavelength form of chlorophyllide [90]:

$$\text{Pchlde } 644/642 \xrightarrow{h\nu} X696 \rightarrow \text{Chlde}684/681 \rightarrow \text{Chlde}674/671$$

The production of nonfluorescent intermediate X696 can proceed below the critical temperature of 200 K, at which all protein motions are frozen out, demonstrating that domain movements or reorganization of the enzyme are not involved at this stage of the catalytic mechanism. The temperature limit for X696 formation is 120 K. These data corroborate with the temperature dependence of X690 production in whole etiolated leaves: when the leaf was frozen to 77 K, only quenching of protochlorophyllide fluorescence was observed upon the formation of the first intermediate R whose spectral features were almost indistinguishable from the spectra of protochlorophyllide [88, 94].

Heyes and co-authors [103] aimed at elucidating the photophysical mechanism of protochlorophyllide reduction in the ternary complexes by comparative study of this reaction at 180 K using absorption spectroscopy, ESR spectroscopy, ENDOR spectroscopy, and Stark spectroscopy. In illuminated samples the ESR spectra were detected suggesting the emergence of two paramagnetic products. However, quantitative estimates based on ESR spectra indicated that the essentially complete transformation of the active protochlorophyllide form (as estimated from changes in absorption spectra) resulted in only 5% output of the pigment free radicals. The appearance of nonfluorescent intermediate with the absorption band at 696 nm corresponded to the broadband Stark effect characteristic of charge transfer steps. The results obtained with Stark spectroscopy provided evidence for the existence of two constituents of the nonfluorescent intermediate. This enabled the authors to suggest that the primary stage of protochlorophyllide photoreduction is associated with the formation of charge-transfer complex. The temperature dependences of the intermediate formation and NADPH oxidation were identical. Therefore, the authors proposed that the formation of nonfluorescent intermediate involves the transfer of hydride ion for the creation of charge-transfer complex. They supposed that the photon absorption by the protochlorophyllide molecule leads to the temporary charge separation along the C17 = C18 double bond, which promotes the ultrafast transfer of hydride ion from NADPH to the C17 atom of protochlorophyllide [103, 105]. The resulting charge transfer complex facilitates the proton transfer toward the C18 atom in the subsequent dark reaction.

The temperature dependence for the rate constants of two sequential reactions (the hydride anion and proton transfer steps) and the respective isotopic effects were measured and analyzed in the form of Eyring plots to obtain thermodynamic parameters for each reaction step [106]. The results of this approach supported the notion that both reactions proceed through the proton tunneling, which requires fast (sub-picosecond) promoting motions coupled to the reaction coordinate. The calculations based on the density functional theory enabled authors to set up a theoretical model of the protochlorophyllide photoreduction; the theory foresees the existence of two potential barriers with activation energies of 166.4 and 147.6 kJ/mol.

The use of pulse-excitation technique provided the opportunity to examine the primary processes in reconstituted active ternary complexes at room temperature [96]. After a 50-fs laser flash (475 nm), a small increase in absorption at 642 nm of the initial protochlorophyllide was observed in the time range between 3 and 400 ps, which was accompanied by a slight shift of the peak position to the short-wavelength spectral region. At the same time, a weak maximum around 677 nm emerged. The short-wavelength intermediate with the absorption bands at 636 and 677 nm was transformed to chlorophyllide.

$$\text{Pchlde} \xrightarrow{h\nu} \text{Pchlde*} \rightarrow \text{Int (636, 677 nm)} \rightarrow \text{Chlde (674 nm)}$$

These results are in accordance with studies on living systems, where, as mentioned above, the primary reaction was also characterized by the lack of protochlorophyllide bleaching (in parallel with effective fluorescence quenching) [95, 107]. The spectral difference between the short-lived intermediate absorbing near 677 nm at room temperature and the intermediate X696 observed at low temperature is probably explained by different conditions of the protein counterpart of the complex at low and high temperatures. The nonfluorescent intermediate X696 is produced upon deep freezing of the complex below the critical temperature of 200 K, known as the temperature limit at which dynamic changes in the enzyme take place. At room temperature the photoreduction of the pigment per se occurs simultaneously with conformational changes in the protein. The importance of flexibility and dynamics in the structure of enzyme for its function was illustrated in the investigations of the protochlorophyllide photoreduction using ultrafast (femtosecond) visible pump-probe absorption spectroscopy and those in the mid-infrared region [108]. The results of measurements allowed the authors to conclude that absorption of the first photon activates the enzyme, which results in a high quantum yield formation of intermediate Int675 on the picosecond timescale when a second photon is absorbed. The authors proposed that minor structural rearrangements, optimizing the alignment of the NADPH-nicotinamide ring and the Tyr residue with the D ring of protochlorophyllide, may be involved in the first light-dependent stage.

Comparative studies of molecular pathways for protochlorophyllide photoreduction *in vivo* and in various model systems allow the conclusion that the primary stages characterized by fluorescence quenching and formation of paramagnetic products are identical for whole etiolated leaves and for all model systems examined, including the most simple systems such as diluted pigment solutions. Nevertheless, some factors suggest that the mechanism of protochlorophyllide photoreduction *in vivo* is somewhat different from the mechanism of the reaction in model systems. First, the active complex configuration *in vivo* seems to be more sophisticated than the ternary complex reconstituted *in vitro* from three main components, i.e., protochlorophyllide, NADPH, and POR. Specifically, this is indicated by different spectral characteristics of artificial and natural active complexes. Etiolated leaves contain several spectrally different active forms of protochlorophyllide. The major forms include Pchld643/639 and Pchld655/650, as well as Pchld653/648 that accumulates and metabolically transforms in juvenile leaves. There are also several minor long-wavelength forms participating in the process of protochlorophyllide photoreduction (reviewed in [109]). The reconstituted ternary active complexes usually comprise one or two protochlorophyllide species: Pchlide633/630 (the form active only at room temperature) and Pchld644/641, which remains photoactive also at low temperatures. Evidence that the mechanism of protochlorophyllide photoreduction *in vivo* is more intricate than *in vitro* was obtained from studies of this process in whole etiolated leaves at low temperature [87,88,93] and from the results of time-resolved spectroscopy at physiological temperatures [100]. These studies demonstrated the multistep nature of the reaction of protochlorophyllide photoreduction, with the existence of two or three short-lived intermediates. Only one short-lived intermediate was observed in the reaction of protochlorophyllide photoreduction in artificial ternary complexes.

The presence of flavins in the active pigment–protein complex and their possible involvement in primary reactions of protochlorophyllide photoreduction was supposed in some studies [110-112]. The results of our studies support this proposal, since illumination of leaves with white or red (>600 nm) light at 77 K was followed by the drop of flavin fluorescence at 520 nm, which was reversed upon the return to higher temperature [112]. In plant leaves flavins might act as an intermediate link of the reaction by accepting hydride ion from NADPH and donating electron to the pigment molecule. In this case the reaction can proceed through a stepwise transfer of two electrons and two protons, leading to the formation at the first step of a semireduced pigment molecule with one attached electron (intermediary semireduced form). This mechanism is largely similar to that known for the reactions of porphyrin photoreduction in solutions [113]. It is not excluded that the whole etiolated leaves possess two mechanisms of hydrogen atom transfer from NADPH to protochlorophyllide: (i) immediate transfer of hydride ion from NADPH to C17 position in the protochlorophyllide molecule with the consequent attachment of proton and/or (ii) flavin-mediated electron transfer from NADPH to the pigment. It is also possible that different mechanisms exist for the primary reactions of spectrally distinct active forms of protochlorophyllide.

Pathways in Formation of Functional Pigment Forms

The terminal photochemical stage of chlorophyll biosynthesis relates not only to photoreduction of the precursor molecule but also to the development of native functional pigment forms incorporated in the photosynthetic apparatus. The existence of several protochlorophyllide species allows the plant to mobilize parallel routes for synthesis of main functional pigment forms attributed to photosynthetic photosystems I and II and to the light-harvesting antenna. Analysis of reactions at the terminal stage of chlorophyll biosynthesis in etiolated, greening, and green leaves by means of low-temperature spectroscopy allowed researchers to elucidate the reaction sequence producing a complex of functional pigment forms.

Continuous efforts of several laboratories have gradually clarified the general sequence and the mechanisms of the reactions involved in the terminal light-dependent stage of chlorophyll production from its precursor, protochlorophyllide in greening etiolated leaves (reviewed in [114]). The outcome of these studies can be summarized with the scheme (Figure 5), which seems to be the most documented and complete to date. The scheme represents a branched tree of the pigment–protein complex conversions producing several native chlorophyll species that constitute the bulk pigment in the light-harvesting antenna and, at the same time, lead to production of minor functionally important pigment forms integrated into the two photosystems of photosynthesis.

Linear reaction pathway (reactions 1–5 in the scheme) includes two consecutive photochemical stages (1 and 2 in the scheme) [115-119] and subsequent dark processes [115-121]. Judging from spectral changes of circular dichroism in homogenates from etiolated plant leaves exposed to illumination, one may suppose that the first photochemical reaction engages one of the two molecules constituting the protochlorophyllide dimer; this reaction produces the complex of two weakly bound molecules (protochlorophyllide and chlorophyllide). The second light reaction performs the photoconversion of the second protochlorophyllide molecule [116-117].

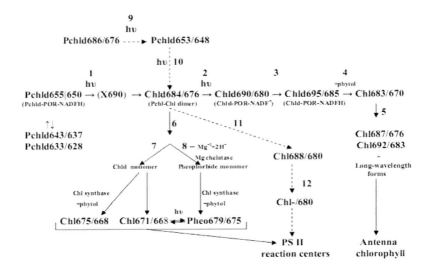

Figure 5. General scheme for conversions of the pigment chromophore at the light stage of chlorophyll biosynthesis in plant leaves. Pchd—protochlorophyllide, Chld—chlorophyllide, Chl—chlorophyll, Pheo—pheophytin. Numbers indicate peak positions of fluorescence (first index) and absorption (second index) for various pigment forms. Solid line arrows designate the pathway leading to formation of native pigments in the light-harvesting complex (reactions 1–5) and the divergent pathway producing pheophytin and two short-wavelength chlorophylls of photosystem II (PSII) reaction centers (path 6–8). Dashed line arrows denote the pathway of producing the pigment P-680 of PSII reaction centers from the long-wavelength form of protochlorophyllide (reactions 9–12). X-690 is a nonfluorescent intermediate.

The temperature-dependent spectral shift Chld690/680 → Chld695/685 (reaction 3 in the scheme)—is light independent and can proceed in darkness [115,120]. This shift reflects the restoration of the reduced hydrogen donor in the pigment–protein complex (reduction of $NADP^+$ produced in the primary photoreaction) [59,119, 122-124]. The results of energy transfer studies led to the suggestion that the bathochromic shift of spectral bands in this reaction can be due to shortening of intermolecular distances in the aggregated active complex [62]. It is possible that the process step considered here is determined by simultaneous action of two factors, i.e., by structural changes of the aggregated complex and by NADPH restoration, provided that the structure of POR complex and the chromophore–chromophore interaction depend on the NADPH redox state (oxidized or reduced).

The reaction 4 in the scheme is manifested as a short-wavelength shift of the pigment spectral bands: Chld 695/685 → Chl683/670 (the Shibata shift) [125,126]. At this stage the chlorophyllide molecule is esterified in plant leaves to produce chlorophyll [127]. At the same time, the disaggregation of the pigment takes place. This is evident from the increase in chlorophyll fluorescence yield [128,129], disappearance of the double signal in circular dichroism spectra [130,131], and disruption of energy transfer from protochlorophyllide to chlorophyll [128,129]. Based on estimated initial distance of about 25 Å between protochlorophyllide molecules within the active complex, Thorne proposed that the disruption of energy transfer implies the increase in the distance between chlorophyll and non-transformed protochlorophyllide by about 10 Å [128]. The pigment disaggregation is apparently due to the disaggregation of the POR dimer [132]. The loosening of the complex is likely to facilitate the pigment esterification, i.e., the attachment of a hydrophobic alcohol to the pigment molecule.

The final stage in the linear chain of reactions (reaction 5 in Figure 5) is manifested in the long-wavelength spectral shift Chl670/683 → Chl676/687; Chl683/692. This shift reflects the final integration of chlorophyll into various pigment–protein complexes of the thylakoid membrane and the formation of chlorophyll spectral forms characteristic of mature green leaves.

The linear reaction pathway is presumably responsible for the production of bulk chlorophyll in the light-harvesting complexes of the two photosystems. Judging from the amount of chlorophyll synthesized by this pathway and its energy coupling (excitation energy transfer) with carotenoids and chlorophyll *b*, this chlorophyll pool is identical to the antenna chlorophyll.

Branching Of Chlorophyll Biosynthetic Pathways: Biosynthesis of Pheophytin A as a Component of Photosystem II Reaction Centers

The intermediate Chld684/676 produced in the first photoreaction (reaction 1 in Figure 5) is the point of divergence of the reaction pathways. Under dark conditions the side reaction of the intermediate conversion into the short-wavelength chlorophyll Chl675/670 (reaction 2a in Figure 5) is observed [115,117-119]. This reaction proceeds at an appreciable rate only at temperatures above 273 K. This reaction can be also observed upon illumination of etiolated leaves with low-intensity light at nonfreezing temperatures. The occurrence of energy transfer from the Chl675/670 species to Chld684/676 indicates that this form is positioned close to the site where the bulk pigment is synthesized [120]. The analysis of pigment extracts by means of thin-layer chromatography showed that the product of the side-path dark reaction is chlorophyll rather than chlorophyllide [115]. Hence, the esterification of chlorophyll molecule occurs not only in association with the "Shibata shift," but it proceeds at a substantially (an order of magnitude) higher rate in the dark reaction of Chl675/670 production from the products of the first photoreaction. The occurrence of two routes for esterification of chlorophyllide in chlorophyll biosynthesis—the fast and slow pathways—was also demonstrated in the study [133]. In the course of this reaction, the pigment is disaggregated [115,116], which is thought to promote its enzymatic esterification. In addition, it was established that the final chlorophyll form produced in the side-path reaction (Chl675/670) is in fact a composition of two chlorophyll forms, Chl671/668 and Chl675/668 [134,135].

The application of low-temperature fluorescence spectroscopy revealed that the side-path reaction sequence not only produces chlorophyll *a* but also gives rise to synthesis of pheophytin *a* (Pheo679/675). This is evidenced from characteristic spectra of fluorescence and fluorescence excitation for illuminated etiolated leaves and pigment extracts [134,135]. The reaction of pheophytin formation represents the biosynthetic process, not the destructive reaction: it proceeds only in intact pigment–protein complexes and is not observed even in homogenates of etiolated leaves. Further studies revealed even a higher complexity of the "side-chain" reaction [135]. It was found out that pheophytin can be transformed to chlorophyll Chl671/668 during the dark reaction and that this reaction is photoreversible.

The complex of pheophytin with chlorophyll in etiolated leaves is in several respects identical to the pheophytin-containing reaction centers of photosystem II (PSII) [134,135].

Biosynthesis of Long-Wavelength Chlorophyll, a Possible Component of PSII Reaction Centers (Reaction Sequence Depicted with Dashed Lines in Figure 5)

The production of long-wavelength chlorophyll, whose spectral characteristics are close to the chlorophyll of PSII reaction center, was observed for the first time upon illumination of heat-shocked etiolated leaves [136]. Thus, a new dark reaction was revealed for the product of protochlorophyllide Chld 684/676 photoreduction, located at the point of the reaction path bifurcation. This reaction is manifested in bathochromic shift of chlorophyllide spectral bands and is accompanied by chlorophyllide esterification: Chld 684/676 → Chl688/680. After the completion of this reaction, fluorescence of the product was entirely quenched within 20–30 s:

Chl 688/680 → Chl–/680. The authors supposed that the final product of this dark reaction chain is the nonfluorescent chlorophyll Chl-680 identical to the pigment P-680 in the PSII reaction center [136].

Investigation of chlorophyll biosynthesis in young (3- to 4-day-old) etiolated leaves yielded additional data concerning the mechanism of chlorophyll P–/680 biosynthesis under natural conditions [137]. In juvenile leaves, like in 7 to 10-day-old leaves grown under heat shock conditions, the intermediate Chlide684/676 participated at room temperature in two dark reactions: Chld 684/676 → Chl 675/670 ("side-path reaction") and Chld 684/676 → Chl688/680 → Chl-/680. A specific feature of the process in young plants was accumulation of Chlide 684/676 within first 3–5 s of white light illumination without appreciable changes in absorption and fluorescence bands of protochlorophyllide Pchlid 655/650, even though the phototransformation of protochlorophyllide (evaluated in extracts) did occur under these conditions. Experiments with the use of monochromatic illumination in the long-wavelength region (680 nm) or white light irradiance at various temperatures led to the conclusion that this phenomenon results from the phototransformation of previously unknown weakly fluorescent long-wavelength protochlorophyllide form Pchlde686/676 into protochlorophyllide Pchld653/648; the latter intermediate converts to chlorophyllide Chld684/676 in the subsequent light reaction.

Thus, at early stages of plant development, apart from the main branched reaction pathway, an additional parallel branched process was found to occur, which synthesizes nonfluorescent chlorophyll Chl–/680 (possibly, the pigment of PSII reaction center) from the long-wavelength protochlorophyllide Pchld686/676 through the intermediate stage of producing the active protochlorophyllide species Pchlide653/648. The intermediate chlorophyllide species Chld684/676 produced by the first photochemical reaction from Pchld655/650 is a triple branching point of biosynthetic pathways leading to the production of minor chlorophyll forms and pheophytin associated with the reaction centers of PSII, as well as to formation of bulk antenna chlorophyll.

In green plant leaves grown under natural photoperiodic conditions, the light-dependent chlorophyll synthesis is implemented with participation of the long-wavelength protochlorophyllide form Pchld654/648, whose spectral characteristics are close to those of the main active form in greening etiolated leaves [59,138]. In analogy to greening leaves, the green leaves possess at least two pathways of chlorophyll biosynthesis, which are recruited depending on the condition and structure of biosynthetic apparatus (prolamellar body and thylakoid formation), which in turn depend on the leaf age and differ during the light and dark periods.

Low-temperature spectroscopic investigations of reconstituted ternary complexes comprising the thermophilic protochlorophyllide oxidoreductase [139] also revealed several intermediary dark reactions in the pathway of conversion from protochlorophyllide Pchld644/642 to chlorophyllide that were related with NADPH oxidation, the release of $NADP^+$ from the complex, subsequent NADPH incorporation, and the final release of free chlorophyllide from the complex and its replacement with a new protochlorophyllide molecule.

Conclusion

Owing to multiyear research of many laboratories, the complex reaction network for the final light-dependent stage of chlorophyll biosynthesis has been clarified. In our view, the main value of the suggested general scheme lies in the fact that it is based on spectroscopic data obtained on intact cells and tissues without any damage to native structures. The research revealed that the multistage conversion of protochlorophyllide to chlorophyll comprises two sequential photoreactions. In addition, this process proceeds in parallel with accomplishing the synthesis of pheophytin that acts as the primary electron acceptor in photochemical system II of photosynthesis. The other simultaneously occurring reaction utilizes a minor long-wavelength protochlorophyllide to synthesize the chlorophyll form that apparently represents the pigment P-680 in reaction centers of photosystem II. The reaction sequence producing the pigment forms of light-harvesting complex has been traced. However, the current scheme of the reaction routes is far from being complete. This particularly concerns the reaction center biogenesis of two photochemical systems of photosynthesis, especially reaction centers of photosystem I.

The mechanisms of intermediary reactions during photoreduction of protochlorophyllide to chlorophyllide remain also the matter of discussion. The photoreaction in etiolated leaves is highly efficient because it proceeds within the photoactive complex assembled prior to photoconversion at the dark stage of chlorophyll precursor synthesis. The active complex comprises protochlorophyllide, NADPH as a hydrogen donor, and the photoenzyme protochlorophyllide oxidoreductase. Owing to specific structure of active ternary complex, particular spatial arrangement of the substrate (protochlorophyllide) and the hydrogen donor NADPH is ensured, which favors the photoreduction process. Both the photoenzyme and protochlorophyllide are likely to occur in the dimer state in etioplasts.

The investigation of protochlorophyllide photoreduction in the intact live cell and in the most elaborate model systems - reconstituted ternary complexes, comprising protochlorophyllide, NADPH, and POR, enabled significant advances in elucidating the mechanism of this reaction. Both in vivo and in model systems perform the photoreduction of chlorophyll precursor through the formation of short-lived free-radical intermediates, with NADPH being a donor of hydride ion and the tyrosine residue of POR serving as a proton donor. However, protochlorophyllide photoreduction in the artificial complex involves only one intermediate, whereas the process in live systems is more complex and includes the formation of two or three short-lived intermediates. Furthermore, the in vivo systems contain a suite of spectrally different forms of chlorophyll precursor that are involved in several parallel reactions significant for creating the pigment system of the photosynthetic apparatus. The subsequent secondary processes occurring in the cell and related with formation of functional chlorophyll forms have not been reproduced *in vitro* thus far. Therefore, we believe that continuation of the research using whole functional cells and tissues is justified and necessary. This approach is especially important for understanding the poorly characterized regulation of final stages in chlorophyll biosynthesis and for practical application of the pathways involved.

References

[1] Griffiths, W. T. (1974). Source of reducing equivalent for the in vitro synthesis of chlorophyll from protochlorophyll. *FEBS Lett*, 46, 301-304.
[2] Griffiths, W. T. (1978). Reconstitution of chlorophyllide formation by isolated etioplast membranes. *Biochem. J.*, 174, 681-692.
[3] Griffiths, W. T. and Mapleston, R. E. (1978). NADPH-protochlorophyllide oxidoreductase. In: G. Akoyunoglou and J.H.Argyroudi-Akoyunoglou (Eds), *Chloroplast Development* (99-104). Amsterdam.
[4] Griffiths, W. T. (1980). Substrate-specificity studies on protochlorophyllide reductase in barlay (Hordeum vulgare) etioplast membranes. *Biochem. J.*, 186, 267-278.

[5] Oliver, R. P. and Griffiths, W. T (1980). Identification of the polypeptides of NADPH: protochlorophyllide oxidoreductase. *Biochem. J.,* 191, 277-280.

[6] Apel, K., Santel, H.J., Redlinger, T.E. and Falk, H. (1980). The protochlorophyllide holochrom of barley (Hordeum vulgare). Isolation and characterization of the NADPH:protochlorophyllide oxidoreductase. *Eur. J. Biochem.,* 111, 251-258.

[7] Baker, M. E. (1994). Protochlorophyllide reduction is homologous to human carbonil reductase and pig 20-beta-hydroxysteroid dehydrogenase. *Biochem. J.,* 300, 605-607.

[8] Wilks, H. M. and Timko, M. P. (1995). A light-dependent complexation system for analysis of NADPH:protochlorophyllide oxidoreductase identification and mutagenesis of two conserved residues that are essential for activity. *Proc. Natl. Acad. Sci. USA*, 92, 724-728.

[9] Birve, S., Selstam, E. and Johansson, L. (1996). Secondary structure of NADPH:protochlorophyllide oxidoreductase examined by circular dichroism and prediction methods. *Biochem. J.,* 317, 549-555.

[10] Suzuki, J.Y. and Bauer, C.E. (1995). A procaryotic origin for light-dependent chlorophyll biosynthesis in plants. *Proc. Natl. Acad. Sci. USA*, 92, 3749-3753.

[11] Fujita, Y. (1996). Protochlorophyllide reduction: a key step in the greening of plants. *Plant Cell Physiol.*, 37, 411-421.

[12] Armstrong, G.A. (1998). Greening in the dark: light-independent chlorophyll biosynthesis fromanoxygenic photosynthetic bacteria to gymnosperms.*J. Photochem. Photobiol.* B, 43, 87-100.

[13] Apel, K. (1981). The protochlorophyllide holochrom of barley (Hordeum vulgare). Phytochrome-induced decrease of translatable mRNA coding for the NADPH:protochlorophyllide oxidoreductase. *Eur. J. Biochem.,* 120, 89-93.

[14] Teakle, R. and Griffiths, W.T. (1993) Cloning, characterization and import studies on protochlorophyllide reductase from wheat (Triticum aestivum). *Biochem. J.,* 296, 225-230.

[15] Townley, H.E., Sessions, R.B., Clarke, A.R., Dafforn. T.R. and Griffiths, W.T. (2001). Protochlorophyllide oxidoreductase: a homology model examined by site-directed mutagenesis. *Proteins: Structure, Funktion, and Genetics,* 44, 329-335.

[16] Wiktorsson, B., Engdahl, S., Zhong, L.B., Boddi, B., Ryberg, M. and Sundqvist C. (1993). The effect of cross-linking of the subunit of

NADFH:protochlorophyllide oxidoreductase on the aggregational state of protochlorophyllide. *Photosynthetica*, 29, 205-218.
[17] Martin, G.E.M., Timko, M.P. and Wilks, H.M. (1997). Purification and kinetic analysis of pea (Pisum sativum L.) NADPY: protochlorophyllide oxidoreductase expressed as fusion with maltose-binding protein in Escherichia coli. *Biochem. J.*, 325, 139-145.
[18] Rowe, J.D. and Griffiths, W.T. (1981). Protochlorophyllide reductase in photosynthetic procaryotes and its role in chlorophyll synthesis. *Biochem. J.*, 311, 417-424.
[19] Schulz, R., Steinmuller, K., Klaas, M., Forreiter, C., Rasmussen, S., Hiller, C. and Apel, A. (1989). Nucleotid sequence of a cDNA coding for the NADPH-protochlorophyllide oxidoreductase (PCR) of barley (Hordeum vulgare L) and its expression in Escherichia coli. *Mol. Gen. Genet.*, 217, 355-361.
[20] Darrah, P.M., Kay, S.A., Teakle, G.R. and Griffiths, W.T. (1990). Cloning and sequencing of protochlorophyllide reductase. *Biochem. J.*, 265, 789-798.
[21] Benli, M., Schulz, R and Apel, K. (1991). Effect of light on the NADPH-protochlorophyllide oxidoreductase of Arabidopsis thaliana. *Plant. Mol. Biol.*, 16, 615-625.
[22] Spano, A. J., He, Z, Michel, H., Hunt, D.F. and Timko M.P. (1992). Molecular cloning, nuclear gene structure, and developmental exprassion of NADPH: protochlorophyllide oxidoreductase in pea (Pisum sativum L.) *Plant. Mol. Biol. V*.18, P. 967-972.
[23] Dahlin C., Sundqvist C and Timko, M.P. (1995). The in vitro assembly of the NADPH-protochlorophyllide oxidoreductase in pea chloroplasts. *Plant. Mol. Biol.*, 29, 317-330.
[24] Klement, H., Helfrich, M., Oster, U., Schoch, S. and Rudiger, W. (1999). Pigment-free NADPH-protochlorophyllide oxidoreductase from Avena sativa L, *Eur. J. Biochem.*, 265, 862-874.
[25] Valera, V., Fung, M., Wessler, A.N. and Richards, W.R. (1987). Synthesis of 4R- and 4S-tritium labeled NADPH for the determination of the coenzyme sterespecificity of NADPH:protochlorophyllide oxidoreductase. *Biochem. Biophy. Res. Comm.*, 148, 515-520.
[26] Begley, J.R. and Young, M. (1989). Protochlorophyllide reductase. I. Determination of the regiochemistry and the stereochemistry of the reduction of protochlorophyllide to chlorophyllide. *J. Am. Chem. Soc.*, 111, 3095-3096.

[27] Lebedev, N., Karginova, O., McIvor, W. and Timko, M. (2001). Tyr275 and Lys279 stabilize NADPH within the catalytic site of NADPH:protochlorophyllide oxidoreductase and are involved in the formation of the enzyme photoactive state. *Biochemistry*, 40, 12562-12574.

[28] Menon, B.R.K., Waltho, J.P., Scrutton, N.S. and Heyes, D.J. (2009) Cryogenic and laser photoexitation studies identify multiple roles for active site residues in the light-driven enzyme protochlorophyllide oxidoreductase. *J. Biol. Chem.*, 284, 18160-18166.

[29] Sytina, O.A., Alexandre, M.T., Heyes, D.J., Hunter, C.N., Robert, B., van Grondelle, R. and Groot, M.L. (2011). Enzyme activation and catalysis: characterization of the vibrational modes of substrate and product in protochlorophyllide oxidoreductase. *Phys. Chem. Chem. Phys.*, 13, 2307 -13.

[30] Schoefs, B. and Franck, F. (2003). Protochlorophyllide reduction: mechanisms and evolution. *Photochem. Photo*biol., 78, 543-557.

[31] Horton, P. and Leech, R.M. (1975). The effect of ATP on the photoconversion of protochlorophyllide in isolated etioplasts of Zea mays. *Plant. Physiol.* 56, 113-120.

[32] Schoch, S., Helfrich, M., Wiktorsson, B., Sundqvist, C., Rudiger, W. and Ryberg, M. (1995). Photoreduction of zinc protopheide b with NADPH-protochlorophyllide oxidoreductase from etiolated wheat (Triticum aestivum L.) *Eur. J. Biochem.*, 229, 291-298.

[33] Griffiths, W.T. (1991). Protochlorophyllide reduction. In: H.,Sheer (Eds), Chlorophyll (433-450). FL, CRC Press, Boca Raton.

[34] Kotzabasis, K., Schuring, M.P. and Senger, H. (1989). Occurrence of protochlorophyll and its transformation to chlorophyll in mutant C-2A' of Scenedesmus obliquus. *Physiol. Plant.*, .75, 221-226.

[35] Ignatov, N. V. and Litvin, F. F. (1996). Photoconversion of long-wavelength protochlorophyll native form Pchl 682/672 into chlorophyll Chl 715/696 in Chlorella vulgaris B-15. *Photosynthes. Res.*, 50, 271-283.

[36] Bjorn, L.O. (1963). Conversion of protochlorophyll in roots. *Physiol. Plant.*, 16, 142-150.

[37] Rebeiz, C.A., Yaghi, M., Abou-Haidar, M. and Castelfranco, P.A. (1970). Protochlorophyll biosynthesis in cucumber (cucumis satius L.) cotyledons. *Plant. Physiol.*, 46, 57-63.

[38] Cohen, C.E. and Schiff, J.A. (1976) Events surrounding the early development of Euglena chloroplast XI. Protochlorophyll (-ide) and its photoconversion. *Photochem. Photobiol.*, .24, 555-566.
[39] Böddi, B., Lindsten, A., Ryberg, M. and Sundqvist, C. (1989). On the aggregational states of protochlorophyllide and its protein complexes in wheat etioplasts. *Physiol. Plant.*, .76, 135-143.
[40] Helfrich, M., Schoch, S., Schafer, W., Ryberg, M. and Rudiger, W. (1996) Absolute configuration of protochlorophyllide a and substrate specificity of NADPH- protochlorophyllide oxidoreductase. *J. Am. Chem. Soc.*, 118, 2606-2611.
[41] Armstrong, G., Runge, S., Frick, G., Sperling, U. and Apel, K. (1995). Identification of NADFH: protochlorophyllide oxidoreductase A and B branched pathway for light-dependent chlorophyll synthesis in Arabidopsis thaliana. *Plant. Physiol.* V.108. P.1505-1517.
[42] Holtorf, H., Reinbothe, S., Reinbothe, C., Bereza, B. and Apel, K. (1995). Two routes of chlorophylle synthesis that are differentially regulated by light in barley (Hordeum vulgare L.). *Proc. Natl. Acad. Sci. USA*, 92, 3254 -3258.
[43] Sperling, U., van Cleve, B., Frick, G., Apel, K. and Armstrong, G.A. (1997). Overexpression of light-dependent PORA or PORB in plants depleted of endogenous POR by far-red light enhances seedlings survival in white light and protects against photooxidative damage. *Plant. J.*, 12, 649-658.
[44] Sperling, U., Franck, F., van Cleve, B., Frick, G., Apel, K. and Armstrong, G.A. (1998). Etioplast differentiation in Arabidopsis: both PORA and PORB restore the prolamellar body and photoactive protochlorophyllide-F655 to the cop1 photomorphogenic mutant. *Plant. Cell V.*10, P. 283-296 .
[45] Franck, F., Sperling, U., Frick, G., Pochert, B., Van Cleve, B., Apel, K and Armstrong, G.A. (2000). Regulation of etioplast pigment-protein complexes, inner membrane architecture, and protochlorophyllide a chemical heterogenety by light-dependent NADPH:protochlorophyllide oxidoreductase A and B. *Plant. Physiol.*, 124, 1678-1696.
[46] Oosawa, N, Masuda, T, Awai, K, Fusada, N, Shimada, H, Ohta, H. and Takamiya, K. (2000). Identification and light-induced expression of a novel gene og NADPH- protochlorophyllide oxidoreductase isoform in Arabidopsis thaliana. *FEBS Lett*, 474, 133-136.
[47] Su, Q, Frick, G, Armstrong, G. and Apel, K. (2001). POR C of Arabidopsis thaliana : a third light- and NADPH-dependent

protochlorophyllide oxidoreductase that is differentially regulated by light. *Plant Molecular Biology*, 47, 805-813.
[48] Frick, G., Su, Q., Apel, K. and Armstrong, G.A. (2003). An Arabidopsis porB porC double mutant lacking light-dependent NADPH:protochlorophyllide oxidoreductase B and C is highly chlorophyll-deficient and developmentally arreste. *Plant. J.*, 35, 141-153.
[49] Santel, H.-J. and Apel, K. (1981). The protochlorophyllide holochrome of barley (Hordeum vulgare L.). The effect of light on the NADFH-protochlorophyllide oxidoreductase. *Eur. J. Biochem.*, 120, 95-103.
[50] Reinbothe, C., Apel, K. and Reinbothe, S. (1995). A light-induced protease from barley plastids degrades NADPH:protochlorophyllide oxidoreductase complexed with chlorophyllide. *Mol. Cell Biol.*, 15, 6206-6212.
[51] Reinbothe, S., Reinbothe, C., Lebedev, N. and Apel, K. (1996). POR A and POR B, two light-dependent protochlorophyllide-reducing enzymes of angiosperm chlorophyll biosynthesis. *Plant Cell*, 8, 763-769.
[52] Masuda, T., Fusada, N., Oosawa, N., Takamatsu, K., Yamamoto, Y.Y., Ohta, M., Nakamura, K., Goto, K., Shibata, D., Shirano, Y., Hayashi, H., Kato, T., Tabata, S., Shimada, H., Ohta, H. and Takamiya, K. (2003). Functional analisis of Isoforms of NADPH: protochlorophyllide oxidoreductase (POR), PORB and PORC, in Arabidopsis thaliana. *Plant Cell Physiol.*, 44, 963-974.
[53] Litvin, F. F. and Stadnichuk, I. N. (1980). Long-wavelength chlorophyll precursors in etiolated leaves and the system of native protochlorophyll forms. *Fiziologiya rastenii*, 27, 1024-1032.
[54] Böddi, B., Ryberg, M. and Sundqvist, C. (1992). Identification of four universal protochlorophyllide forms in dark-grown leaves by analyses of the 77 K fluorescence emission spectra. *J. Photochem. Photobiol.*, 12, 389-401.
[55] Schoefs, B., Bertrand, M. and Franck, F. (2000). Spectroscopic properties of protochlorophyllide analized in situ in the course of etiolation and in illuminated leaves. *Photochem. Photobiol.*,72, 85-93.
[56] Stadnichuk, I. N., Amirjani, M. R. and Sundqvist, C. (2005). Identification of spectral forms of protochlorophyllide in the region 670-730 nm. *Photochem. Photobiol. Sci.*, 4, 230-238.
[57] Ignatov, N. V. and Litvin, F. F. (2002). A new pathway of chlorophyll biosynthesis from long-wavelength protochlorophyllide Pchlide 686/676 in juvenile etiolated plants. *Photosynth. Res.*, 71, 195-207.

[58] Franck, F., Berthelemy, X. and Strazlka, K. (1993). Spectroscopic characterization of protochlorophyllide reduction in the greening leaf. *Photosyinthetica*, 29, 185-194.

[59] Schoefs, B. and Franck, F. (2008). The photoenzymatic cycle of NADPH: protochlorophyllide oxidoreductase in primary bean leaves (Phaseolus vulgaris) during the first days of photoperiodic growth. *Photosynth. Res.*, 96, 15-26.

[60] Kahn, A., Boardman, N. K. and Thorne, S. W. (1970). Energy transfer between protochlorophyllide molecules: evidence for multiple chromophores in the photoactive protochlorophyllide-protein complex in vivo and in vitro. *J. Mol. Biol.*, 48, 85-101.

[61] Litvin, F. F., Efimtsev, E. I. and Ignatov, N. V. (1976). Energy transfer in photoactive complexes of chlorophyll precursor in etiolated leaves and spectroscopic characteristics of pigment forms. *Biofizika*, 21, 307-312.

[62] Ignatov, N. V. and Litvin, F. F. (1981). Energy transfer in pigment complexes of protochlorophyllide. *Biofizika*, 26, 664-668.

[63] Akulovich, N. K. and Orlovskaya, K. I. (1971). Spectral characterization of protochlorophyll(ide)–holochrome during its production and transformation to chlorophyll in etiolated plants of various types. In: Metabolism and structure of photosynthetic apparatus (3-21). Minsk: *Nauka i tekhnika*.

[64] Valter, G., Belyaeva, O. B., Ignatov, N. V., Krasnovsky, A. A. and Litvin, F. F. (1982). Photoconversions of various protochlorophyll(ide) forms in Phaseolus coccineus. *Biologicheskie nauki*, no. 9, 35-39.

[65] Houssier, C. and Sauer, K. (1970). Circular cichroism and magnetic circular cichroism of the chlorophyll and protochlorophyll pigments. *J. Amer. Chem. Soc.*, 92, 779-791.

[66] Vaugan, G. D. and Sauer, K. (1974). Energy transfer from protochlorophyllide to chlorophyllide during photoconvertion of etiolated bean holochrome. *Biochem. Biophys. Acta*, 347, 383-394.

[67] Bovey, F., Ogawa, T. and Shibata, K. (1974). Photoconvertible and non-photoconvertible forms of protochlorophyll(ide) in etiolated bean leaves. *Plant Cell Physiol.*, 15, 1133-1137.

[68] Shioi, Y. and Sasa, T. (1984). Chlorophyll formation in the YG-6 mutant of Chlorella vulgaris: spectral characterization of protochlorophyllide phototransformation. *Plant Cell Physiol.*, 25, 131-137.

[69] Franck, F. and Strazlka, K. (1992). Detection of the photoactive protochlorophyllide-protein complex in the light during the greening of barley. *FEBS Lett.*, 309, 73-77.
[70] Mysliwa-Kurdziel, B., Amirjani, M. R., Strzalka, K. and Sundqvist, C. (2003). Fluorescence lifetime of protochlorophyllide in plants with different proportions of short-wavelength and long-wavelength protochlorophyllide spectral forms. *Photochem. Photobiol.*, 78, 205-212.
[71] Griffiths, W. T. (1975). Characterization of the terminal stages of chlorophyll(ide) synthesis in etioplast membrane preparations. *Biochem. J.*, 152, 623-635.
[72] Sundqvist C. and Dahlin C. (1997). With chlorophyll from prolamellar bodies to light-harvesting complexes. *Physiol. Plant* V.100, 748-759.
[73] Schoefs, B. and Franck, F. (1993). Photoreduction of protochlorophyllide to chlorophyllide in 2-day old dark-grown bean leaves. Comparison with 10-day old leaves. *J. Exp. Bot.*, 44, 1053-1057.
[74] Dubrovsky, V. T. and Litvin, F. F. (2008). Detection of early photoactive forms of chlorophyll precursor in plant leaves by means of fluorescence spectroscopy at 20°C. *Biologicheskie membrany*, 25, 203-209.
[75] Bystrova M. I., Lang F. and Krasnovsky, A. A. (1972). Spectral effects of protochlorophyllous pigment aggregation. *Molekulyarnaya Biologiya*, 6, 77-86.
[76] Brouers, M. (1972). Optical properties of in vivo aggregates of protochlorophyllide in non-polar solvents. I. Visible and fluorescence spectra. *Photosynthetica*, 6, 415-423.
[77] Böddi, B., Soos, J. and Lang, F. (1980). Protochlorophyll forms with different molecular arrangements. *Biochim. Biophys. Acta*, 593, 158-165.
[78] Sundqvist, C., Ryberg, H., Boddi, B. and Lang, F. (1980). Spectral properties of a long-wavelength absorbing form of protochlorophyll in seeds of Cyclantera explodens. *Physiol. Plant*, 48, 297-301.
[79] Ignatov, N. V., Belyaeva, O. B., Timofeev, K. N. and Litvin, F. F. (1988). Photochemical reactions of protochlorophyll in the inner seed coat layers of Cucurbita. *Biofizika*, 33, 500-505.
[80] Rubin, A.B., Minchenkova, L.E. Krasnovsky, A.A. and Tumerman, L.A. (1962) .Investigation of protochlorophyllide fluorescence lifetime during greening of etiolated leaves. *Biofizika* (Moscow), 7, 571-577.

[81] Goedheer J.C. and Verhulsdonk C.A.H. (1970). Fluorescence and phototransformation of protochlorophyll with etiolated bean leaves from -196°C to +20°C. *Biochem. Biophys. Res. Comm.*, 39, 260-266.
[82] Sironval, C. and Kuyper, P. (1972). The reduction of protochlorophyllide into chlorophyllide: IV. The nature of the intermediate P688-676 species. *Photosynthetica*, 6, 254-275.
[83] Dujardin, E. and Sironval, C. (1977). The primary reactions in the protochlorophyll(ide) photoreduction. *Plant Science Letters*, 10, 347-353.
[84] Dujardin, E. and Correia, M. (1979). Long-wavelength absorbing pigment protein complexes as fluorescence quenchers in etiolated leaves illuminated in liquid nitrogen. *Photobiochem. Photobiophys.*, 1, 25-32.
[85] Dujardin, E. (1984). The long-wavelength-absorbing quenchers formed during illumination of protochlorophyllide-proteins. In: C., Sironval and M., Brouers, (Eds.), *Protochlorophyllide Reduction and Greening* (pp. 99-112). The Hague, Martinus Nijhoff/Dr.W Junk Publisher.
[86] Losev, A.P. and Lyal′kova, N.D. (1979). Investigation of the primaries stages of protochlorophyllide photoreduction in the etiolated plants. *Molecular Biology* (Moscow), 13, 837-844.
[87] Belyaeva O.B. and Litvin, F.F. (1981). Primary reactions of protochlorophyllide into chlorophyllide phototransformation at 77 K. *Photosynthetica*, 15, 210-215.
[88] Litvin, F.F., Ignatov, N.V. and Belyaeva, O. B. (1981). Photoreversibility of transformation of protochlorophyllide into chlorophyllide. *Photobiochem. Photobiophys.*, 2, 233-237.
[89] Heyes, D.J., Ruban, A.V., Wilks, H.M. and Hunter, C.N. (2002). Enzimology below 200 K: The kinetics and thermodinamics of the photochemistry catalyzed by protochlorophyllide oxidoreductase. *Proc. Nat. Acad. Sci. USA*, 99, 11145-11150.
[90] Heyes, D.J. Ruban A.V. and Hunter, C.N. (2003). Protochlorophyllide oxidoreductase: "Dark" reaction of a light-driven enzyme. *Biochemistry*, 42, 523-528.
[91] Raskin, V.I. (1976) Mechanism of photoredution of protochlorophyllide in the intact etiolated leaves. *Vesti Akademii Nauk BSSR*, no. 5, 43-46.
[92] Belyaeva, O.B. Personova, E.R. Litvin, F.F. (1983). Photochemical reaction of chlorophyll biosynthesis at 4,2 K. *Photosynthesis Research*, 4, 81 - 85.

[93] Belyaeva, O.B. Timofeev K.N. and Litvin, F.F. (1988). The primary reaction in the protochlorophyll(ide) photoreduction as investigated by optical and ESR-spectroscopy. *Photosynthesis Research*, 15, 247-256.
[94] Belyaeva, O.B. (2009). *Light dependent chlorophyll biosynthesis.* Moscow, BINOM.
[95] Iwai, J., Ikeuchi, M., Inoue, Y. and Kobayashi, T. (1984). Early processes of protochlorophyllide photoreduction as measured by nanosecond and picosecond spectrophotometry. In: C., Sironv and M., Brouers, (Eds.), *Protochlorophyllide Reduction and Greening* (99-112). The Hague, Martinus Nijhoff/Dr.W Junk Publisher.
[96] Heyes, D.J., Hunter, C.N., van Stokkum, I.H.M., Grondelle, R. and Groot, M.L. (2003). Ultrafast enzymatic reaction dynamics in protochlorophyllide oxidoreductase. *Nature Structural Biology*, 10, 491-492.
[97] Dobek, A., Dujardin, E., Franck, F., Sironval, C., Breton, J. and Roux, E. (1981). The first events of protochlorophyll(ide) photoreduction investigated in etiolated leaves by means of the fluorescence excited by short, 610 nm laser flashes at room temperature. *Photobiochem. Photobiophys.*, 2, 35-44.
[98] Belyaeva, O.B. and Sundqvist, C. (1998). Comparative investigation of the appearance of primary chlorophyllide forms in etiolated leaves, prolamellar bodies and prothylakoids. *Photosynth. Research*, 55, 41-48.
[99] Frank, F., Dujardin, E. and Sironval, C. (1980). Non-fluorescent, short-lived intermediate in photoenzymatic protochlorophyllide reduction at room temperature. *Plant Sci. Lett.*, 18, 375-380.
[100] Inoue, Y., Kobayashi, T., Ogawa T., and Shibata, K. (1981). Ashort intermediate in the photoconversion of protochlorophyllide to chlorophyllide a. *Plant Cell Physiology*, 22, 97-204.
[101] Franck F. and Mathis, P. (1980). A short-lived intermediate in the photoenzimatic reduction of protochlorophyll(ide) into chlorophyll(ide) at a physiological temperature. *Photochem. Photobiol.*, 32, 799-803.
[102] Lebedev N. and Timko M. (1999). Protochlorophyllide oxidoreductase B-catalyzed protochlorophyllide photoreduction in vitro: Insigh t into the mechanism of chlorophyll formation in light-adapted plants. *Proc. Natl. Acad. Sci. USA*, 96, 17954-17959.
[103] Heyes D.J., Heathcote P., Rigby S.E.J., Palacios M.A., Grondelle R. and Hunter C.N. (2006). The first catalytic step of the light-driven enzyme protochlorophyllide oxidoreductase proceeds via a charge transfer complex. *J. Biol. Chem.*, 281, 26847-26853.

[104] Belyaeva, O.B., Griffiths, W.T., Kovalev, J.V., Timofeev, K.N. and and Litvin, F.F. (2001). Participation of free radicals in photoreduction of protochlorophyllide to chlorophyllide in artificial pigment–protein complexes. *Biochemistry* (Moscow), 66, 173-177.

[105] Griffiths W.T., McHugh T. and Blankenship R.E. (1996). The light intensity dependence of protochlorophyllide photoreduction and its significance to the catalytic mechanism of protochlorophyllide reductase. *FEBS Lett.*, 398, 235-238.

[106] Heyes, D.J., Sakuma, M., Visser S.P. and Scrutton, N.S. (2009). Nuclear quantum tunneling in the light-activated enzyme protochlorophyllide oxidoreductase. *J. boil. Chem.*, 284, 3762-3767.

[107] Ignatov, N., Belyaeva, O. and Litvin, F. (1993). Low temperature phototransformation of protochlorophyll(ide) in etiolated leaves. *Photosynth. Res.*, 38, 117-124.

[108] Sytina, O.A., Heyes, D.J., Hunter, C.N., Alexandre, M.T., van Stokkum, I.H.M., van Grondelle, R. and Groot, M.L. (2008). Conformational changes in an ultrafast light-driven enzyme determine catalytic activity. *Nature*, 456, 1001-1005.

[109] Belyaeva, O.B. and Litvin, F.F. (2007). Photoactive pigment-enzyme complexes of chlorophyll precursor. *Biochemistry* (Moscow), 72, 1458-1477.

[110] Walker, C.J. and Griffiths, W.T. (1988). Protochlorophyllide reductase: a flavoprotein?. *FEBS Letters*, 239, 259-262.

[111] Nayar, P., Brun, A., Harriman, A. and Begley, T.P. (1992). Mechanistic studies on protochlorophyllide reductase: a model system for the enzymatic reaction. *J. Chem. Soc. Chem. Comm.*, issue 5, 395-397.

[112] Ignatov, N.V., Belyaeva, O.B. and and Litvin, F.F. (1993). The possible role of the flavin components of protochlorophylide-protein complexes in the primary processes of protochlorophyll photoreduction in etiolated plant leaves. *Photosynthetica*, 29, 235-241.

[113] Terenin, A.N. (1967). Molecular Photonics of Dyes and Related Organic Compounds. Leningrad, *Nauka*.

[114] Belyaeva, O.B. and Litvin, F.F. (2009). Pathways of formation of pigment forms at the terminal photobiochemical stage of chlorophyll biosynthesis. *Biochemistry,* 74, 1535 -1544.

[115] Litvin, F.F. and Belyaeva, O.B. (1971). Sequence of photochemical and dark reactions in the terminal stage of chlorophyll biosynthesis. *Photosynthetica,* 5, 200-209.

[116] Mathis, P. and Sauer, K. (1972). Circular dichroism studies on the structure and the photochemistry of protochlorophyllide and chlorophylllide holochrome. *Biochim. Biophys. acta*, 267, 498-511.
[117] Mathis, P. and Sauer, K. (1973). Chlorophyll formationin greening bean leaves during theearly stages. *Plant Physiol.,* 51, 115-119.
[118] Litvin, F., Ignatov, N., Efimtsev, E. and Belyaeva, O. (1978). Two successive photocemical reactions in protochlorophyll(ide) reduction into chlorophyll(ide). *Photosynthetica*, 12, 375-381.
[119] Oliver, R.P. and Griffiths, W.T (1982). Pigment-protein complexes of illuminated etiolated leaves. *Plant Physio.*, 70, 1019-1025.
[120] Sironval, C., Kuyper, Y., Michel, J.M. and Brouers, M. (1967). The primary photoact in the conversion of protochlorophyll into chlorophyllide. *Stud. Biophys.*, 5, 43-50.
[121] Henningsen, K.W. and Thorne, S.W. (1974). Esterification and spectral shift of chlorophyll (ide) in wild-type and mutant seedlings development in darknees. *Physiol. Plat.*, 30, 82-89.
[122] El Hamouri, B. and Sironval, C. (1980). NADP+/NADPH control of the protochlorophyllide-chlorophyllide proteins in cucumber etioplasts. *Photobiochem. Photobiophys.*, 1, 219-223.
[123] El Hamouri, B. and Sironval, C. (1981). Pathway from photoinactive P633-628 protochlorophyllide to the P696-682 chlorophyllide in cucumber etioplast suspensions. *Plant Science Lett.*, 21, 375-379.
[124] Franck, F., Bereza, B. and Boddi, B. (1999). Protochlorophyllide-NADP+ and protochlorophyllide-NADPH complexes and their regeneration after flash illumination in leaves and etioplast membranes of dark-grown wheat. *Photosynth. Res.*, 59, 53-61.
[125] Shibata, K. (1956). Spectoscopic studies on chlorophyll formation in intact leaves. *Carn.Inst.Wash. YB*, 55, 248-250.
[126] Shibata, K. (1957). Spectoscopic studies on chlorophyll formation in intact leaves. *J. Biochem.*(Tokyo), 44, 147-173.
[127] Henningsen, K. W. and Thorne, S. W. (1974). Esterification and spectral shift of chlorophyll (ide) in wild-type and mutant seedlings development in darknees. *Physiol. Plant*, 30, 82-89.
[128] Thorne, S. W. (1971). The greening of etiolated bean leaves. II. Secondary and further photoconversion processes. *Biochem. Biophys. Acta,* 226, 128-134.
[129] Thorne, S. W. (1971). The greening of etiolated bean leaves. III.Multiple light/dark step photoconversion processes. *Biochem. Biophys. Acta*, 226, 175 -183.

[130] Henningsen, K. W., Kahn, A. and Houssier, C. (1973). Circular dichroism of protochlorophyllide and chlorophyllide holochrome subunits. *FEBS Lett.*, 37, 103-109.
[131] Schultz, A. and Sauer, K. (1972). Circular dichroism and fluorescence changes accompaning the protochlorophyllide to chlorophyllide transformation in greening leaves and holochrome preparations. *Biochem. Biophys. Acta*, 267, 320-340.
[132] Wiktorsson, B., Ryberg, M., Gough, S. and Sundqvist, C. (1992). Isoelectric focusing of pigment-protein complexes solubilized from non-irradiated and irradiated prolamellar bodies. *Physiol. Plant*, 85, 659-669.
[133] Domanski, V., Rassadina, V., Gus-Mayer, S., Wanner, G., Schoch, S. and Rudiger, W. (2003). Characterization of two phases of chlorophyll formation during greening of etiolated barley leaves. *Planta*, 216, 475-483.
[134] Ignatov, N.V. and Litvin, F.F. (1994). Photoinduced formation of pheophytin/chlorophyll-containing complexes during the greening of plant leaves. *Photosynthes. Res*, 42, 27-35.
[135] Ignatov, N.V. and Litvin, F.F. (1995). Light-regulated pigment interconversion in pheophytin/ chlorophyll-containing complexes formed during plant leaves greening. *Photosynthes. Res.* V.46, P.445-453.

Ignatov, N.V., Satina L.Y. and Litvin, F.F. (1999). Biosynthesis of non-fluorescent chlorophyll of photosystem II core in greening plant leaves. Effect of etiolated plants growing under heat shock conditions (38° C). *Photosynth. Res.*, 62, 185-195.
[136] Ignatov, N.V. and Litvin, F.F. (2002). A new pathway of chlorophyll biosynthesis from long-wavelength protochlorophyllide Pchlide 686/676 in juvenile etiolated plants. *Photosynth. Res.*, 71, 195-207.
[137] Ignatov, N. V. and Litvin, F. F. (2002). Chlorophyll biosynthesis from protochlorophyllide in green plant leaves. *Biokhimiya*, 67, 1142-1150.
[138] Heyes D.J. and Hunter C.N. (2004). Identification and characterization of the product release steps within the catalytic cycle of protochlorophyllide oxidoreductase. *Biochemistry,* V.43, P.8265-8271.

In: Chlorophyll
Editors: H. Le, et.al.

ISBN: 978-1-61470-974-9
© 2012 Nova Science Publishers, Inc.

Chapter III

Genomics and Phenomics of Chlorophyll Associated Traits in Abiotic Stress Tolerance Breeding

U. R. Rosyara,[1*] *N. K. Gupta,*[2] *S. Gupta,*[2] *and R. C. Sharma*[3]

[1]Michigan State University, Crop and Soil Science Department, East Lansing, MI, U. S.
[2]Rajasthan Agricultural University, Plant Physiology Department, Bikaner, India
[3]International Center for Agricultural Research in the Dry Areas (ICARDA), Central Asia and the Caucasus Regional Program, Tashkent, Uzbekistan

Abstract

Development of stress tolerance in economically important crops is very important in context of recent challenges brought by changing climate along with increased demand for both food and fuel. Crop plants

---

[*] Corresponding author: rosyara@msu.edu.

are affected by several abiotic factors such as high or low temperature, excessive water or drought conditions, low or high light, nutrient deficiency or nutrient excess, salt, pollutant and heavy metal. Studies have shown genetic variation in tolerance to these stresses and chlorophyll associated traits such as stay green, chlorophyll content, chlorophyll fluorescence and spectral reflectance are considered as indicator to stress tolerance to one or more of these stresses. Advancement in science of abiotic stress tolerance breeding relies on advancement in genomics and phenomics. The phenomics of stress tolerance are now focused on development of precision high throughput techniques whereas genomics has advanced our knowledge on genes or quantitative trait loci (QTL) allowing selection to perform at gene level and production of genetically engineered crops. This review article presents an insight into practical application of physiology and molecular biology of chlorophyll alternation in response to stresses in context of development of novel stress tolerant plant genotypes.

Keywords: stress tolerance breeding, genomics, phenomics, chlorophyll content, chlorophyll fluorescence, physiological breeding, molecular breeding

1. Introduction

World food production is restricted principally by biotic and abiotic stresses. Abiotic stresses predominately caused by environment are considered principal factor causing more than 70% yield reductions (Boyer et al., 1982). Every types of abiotic stresses whether it is water deficiency or excess, low or high temperature, nutrient deficiencies or nutrient excesses or salinity or mineral toxicities, low or high light reduce plant growth impacting on crop yield (Cramer 2010; Langridge et al. 2006; Munns and Tester, 2008; Witcombe et al., 2008; Salekdeh, et al. 2009; Rosyara et al. 2010a,b; 2009, 2008a,b; 2007a,b; 2006; Shabala et al., 1998; Reynolds et al., 2001). The stresses are becoming more pertinent in today's and future environment due to climate change, land degradation, land and water pollution (Tester and Langridge, 2010; Witcombe, 2008; Reynolds et al., 2001). This only shows relevance of abiotic stresses, however, does not undermine relevance of biotic stresses (Boyer, 1982) and sometime they can be interacting or confounded with biotic stresses (Rosyara et al. 2010a,b; 2009, 2008a,b). Nevertheless, in this chapter we will primarily focus on abiotic stresses with prime focus on chlorophyll associated traits. Development of stress tolerant genotypes is one of the most important research objectives to combat abiotic stresses in current and future environments. In this regard precision genomics and phenomics along with different molecular breeding tools are key instrument to meet immense challenges in improvement of crops in future.

2. Physiology and Genomics of Chlorophyll Processes

2.1. The Chlorophyll

The chlorophyll molecule, also known as photoreceptor, is a central player in harvesting light energy for photosynthesis (Salisbury and Ross, 1986). Functionally, the chlorophyll is part of the multiple protein–pigment photosynthetic complexes that capture light photon, a first step in converting light energy into chemical energy and carbon assimilation. The fundamental structure of a chlorophyll molecule is a porphyrin ring, coordinated to a central atom (Salisbury and Ross, 1986). There are two main types of chlorophyll molecules: Chlorophyll a and Chlorophyll b. These two differ only slightly, in the composition of a side chain (in a it is -CH3, in b it is CHO) and color: Chlorophyll a is blue whereas Chlorophyll b is yellow green (Salisbury and Ross, 1986). Both of these two chlorophylls are very effective photoreceptors with strong absorption bands in the visible regions of the spectrum (390 to 750 nm). Chlorophyll molecules are specifically arranged in and around photosystems (Photosystem II with reaction center P680 and Photosystem I with reaction center P700) that are embedded in the thylakoid membranes of chloroplasts. The physiology of photosynthesis and chlorophyll metabolism has been well established along with development in the areas of genetics and genomics, genes responsible for the metabolism have also been discovered.

2.2. Genomics of Chlorophyll Metabolism

The chlorophyll physiology (i.e. metabolism, synthesis and degradation) in relation to photosynthesis has been area of interest for researcher with objective of development of stress tolerance plants. With advances in knowledge of genomics or genetics of key physiological processes, chlorophyll metabolism is now more understood, although is not fully understood yet.

The biosynthetic chain of chlorophyll commences with the tiny building blocks, acetate and glycine molecules that are part of basic metabolic environment. These small molecules further condense into a series of n steps to thereby forming the complex molecule protoporphyrin (Granick, 1951). Isoprenoid biosynthetic pathway is responsible for providing intermediates for the synthesis of chlorophyll (Lange and Ghassemian, 2003). Isoprenoids constitute a huge family of compounds with more than 20,000 members (Chappel 1995). The cytosolic mevalonate (MVA) independent pathway is responsible for synthesis of precursors involved in the biosynthesis side chains of chlorophylls (Eisenreich et al., 2001). Interestingly, outside of the photosynthetic complex, the photodynamic properties of chlorophyll result in phototoxicity of its anabolic and catabolic metabolites. Since biosynthesis and breakdown of chlorophyll occur throughout plant development process, in order to protect tissues from accumulation of intermediate compounds that can potentially damage tissues, these pathways must be tightly regulated to prevent. Lange and Ghassemian (2003) have presented a comprehensive analysis of genes coding for enzymes involved in the metabolism of isoprenoid-derived compounds in *Arabidopsis thaliana.*

Beside synthesis of chlorophyll breakdown or catbolism is very important in perspective of natural senescence as well as stress tolerance. Chlorophyll catabolism is dramatically visualized during leaf senescence (Goldschmidt, 2001) which is moderated by environmental conditions such as excessive light, leading to photoinhibition (Prasil et al., 1992; Andersson and Barber, 1996).

Insight into the chlorophyll catabolic pathways has been provided by characterization of chlorophyll breakdown products in plant tissues (Hörtensteiner, 1999, 2006; Amir-Shapira et al., 1987; Engel et al., 1991; Matile et al., 1996). The breakdown of chlorophyll into magnesium, phytol, and the primary cleavage product of the porphyrin ring happens in four consecutive steps. These steps are catalyzed by chlorophyllase (Chlase), Mg-dechelatase, pheophorbide *a* oxygenase, and red chlorophyll catabolite reductase. The genomic studies have characterized some of responsible genes for the catabolism in different plants. Genes for the majority of the steps from the chlorophyll catabolism pathway have already been cloned (Gray et al., 2002; Jakob-Wilk et al., 1999; Pruiinska et al., 2003; Tanaka et al., 2003; Tommasini et al., 1998; Tsuchiya et al., 1999; Wiithrich et al., 2000). The genes Chlorophyllase from Citrus and Chenopodium (Jakob-Wilk et al. 1999;

Tsuchiya et al. 1999), and Pheophorbide, a Oxygenase that was shown to correspond to the accelerated *cell death1* gene in *Arabidopsis* (Pruzinska et al. 2003) are examples of the genes reported.

A number of genetic variants, mutations, and transgenics altering chlorophyll catabolism have been described (Pruiinska et al., 2003; Thomas and Howarth, 2000; Thomas et al., 2001). Broadly, they fall into two main categories:

a. Stay-greens with slower senescence (Thomas and Smart, 1993).
b. Photosensitive genotypes variants with deficiencies in particular steps of tetrapyrrole metabolism display pathological symptoms that often mimic disease lesions and are consistent with the accumulation of photodynamic intermediates upstream of the metabolic blockage (Hortensteiner, 2004).

Studies have reported some specific mutants responsible for reduced chlorophyll loss or maintaining chlorophyll fluorescence under stress conditions. Some of the example includes the gene *"adc"* responsible for Arginine decarboxylase (Capell et al., 1998) and the gene *"TPSP"* responsible for Trehalose synthesis (In-Cheol Jang et al., 2003). The groups of osmoprotectants are also of particular interest for improving abiotic stress tolerance of crops. The trehalose synthase (*TSase*) gene of the edible wood fungi *Grifola frondosa* has shown increased tolerance to drought and salt indicated by physiological indicator of stress tolerance including chlorophyll *a* and *b* contents (Zhang et al. 2005, 2006). Also some genes have been discovered that inhibit chlorophyll degradation process for example, *chlorophyll retainer* (*cl*) mutation in pepper (Efrati et al. 2005).

2.3. Quantitative Trait Loci or Major Genes and Stress Tolerance

In majority of instances the stress tolerance is measure quantitatively in term of chlorophyll associated traits by exploiting difference between stress and non stressed situation. Such quantitative variation in the trait has been genetically diagnosed in studies involving Quantitative trait loci (QTLs), QTL have been identified in numerous wheat, barley and rice populations for heat(Mason et al. 2010, Mohammadi et al. 2008, Yang et al. 2002), drought (Chen et al. 2010, Mathews et al. 2008, Quarrie et al. 2006, Kirigwi et al.

2007, von Korff et al. 2008, Yang et al. 2007, Peleg et al 2009), cold (, Andaya et al. 2006, Baga et al. 2007; Kuroki et al. 2007), nutrient deficiencies (Laperche et al. 2008, Yang et al. 2010, Wissuwa et al. 2006), salinity(Prasad et al., 2000; Mano et al. 1997, Ma et al., 2007; Lee et al. 2007, Zhang et al. 1995, Lin et al. 2004), and mineral toxicity (Balint et al. 2007, 2009; Cai et al. 2008, Ma et al. 2006, Navakode et al. 2009). Different approaches make use of genetic diversity responsible adoption in different environmental conditions both in landraces as well as wild varieties to identify tolerance genes (Balint et al. 2007, 2009). A large number of genes with a potential role in stress tolerance (Ingram and Bartels, 1996; Flowers and Yeo, 1995; Hall 2002; Langridge et al., 2006; Collins et al., 2008; Genc et al., 2010; and Fleury et al., 2010). The comprehensive reviews by Hall (2002), Langridge et al. (2006), Collins et al. (2008) Genc et al. (2010) and Fleury et al. (2010) provide a more comprehensive coverage of QTL linked to abiotic stress tolerance in different crop plant species.

3. Stress and Chlorophyll Associated Traits

Different physiological processes inside plants have effect of chlorophyll metabolism in plants and consequently effect photosynthesis processes. Both high and low temperatures are reported to inhibit biosynthesis of chlorophyll (Tewari and Tripathy, 1998). Active oxygen species (AOS) are for all time associated with aerobic life (Vranova et al., 2002). Abiotic stresses such as water stress, temperature stress, light stress, salt stresses, nutrient stress, heavy metal stress and pollution stress are known to accelerate the production of AOS in plants that result in damage to membrane systems and other cellular processes (Dat et al., 2000; Mittler, 2002; Mittler et al., 2004). Antioxidative systems, both enzymatic and non-enzymatic systems, play an important role in balancing and preventing oxidative damage (Bowler et al., 1994; Foyer et al., 1994). However, the production and efficiency of the antioxidative systems depend on plant type and genetic makeup of the plant. Such AOS are able to impact chlorophyll biosynthesis (Aarti et al. 2006). Stress tolerance usually involves different genetically controlled physiological processes, the outcome can be measured as chlorophyll related traits including chlorophyll content, chlorophyll fluorescence and stay green (Figure 1) Obviously different types of stress have effect on chlorophyll properties in leaves and other plant parts, has been considered important indicator for stress tolerance breeding (Reynolds et al. 2001). In following section, we will discuss measurement of chlorophyll associated traits (Figure 2) with special focus on breeding implications.

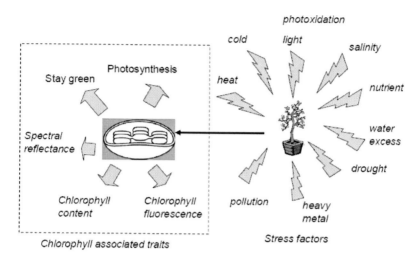

Figure 1. Major stress factors and important chlorophyll related properties.

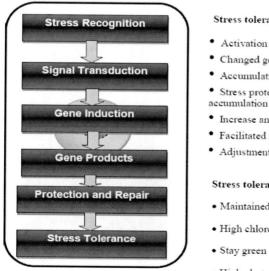

Figure 2. Stress tolerance mechanism and important phenotypic indicators.

3.1. Phenomics of Chlorophyll Associated Traits

Phenomics is important area of biology related to measurement of phenomes i.e. physical and biochemical traits, of organisms in response to genetic and environmental effects. Precise measurement of phenotypes is one of important activity for successful genetics and breeding programs. For plant breeding in addition to be precise, less affected by measurement errors or environments should be ideally using rapid, non-invasive technique. For effective molecular studies precise phenotypic is very important. The following criteria are considered before using any trait in a breeding program:

(a) Degree of traits association
(b) Ease of measurement
(c) Precision in measurement
(d) Heritability of trait and response to selection

The genetic response of using any trait in indirect selection we should consider correlated response (Falconer and Mackay, 1996)

$$CR = i.h_x. r_g. \sigma_{gy} \qquad (1)$$

CR is the correlated response for the primary trait (for example yield under stressed condition), from selection based on character x; i is the standardized selection differential (related to the proportion of genetic population selected); h_x is the square root of the heritability for character x; r_g is the genetic correlation between character y and character x; and σ_{gy} is the genetic standard deviation for character y.

Figure 3. Decision triangle using direct or indirect selection.

As illustrated in figure 3, heritability of trait is subject to change depending upon the control over the environmental variation, molecular markers are considered best for their heritability but should have high co-heritability for using in a breeding program along with technological cost of measurement. Physiological traits on other hand are modest in cost of measurement however the heritability and co-heritability is modest in some conditions. In following sections we will discuss on some promising chlorophyll associated traits useful in stress tolerance breeding perspective.

3.2. Genotyping or Phenotyping or 'Physiotyping'?

Characterization for abiotic stress tolerance is complex, but improved phenotyping technologies can help in precise quantification of abiotic stress quickly, economically and practically, thereby fitting into day to day breeding program. In parallel with advances in precision phenomics, genomics is becoming advanced with development of sequencing and resequencing technology. Also such information is more publicly available. Thus, genetic knowledge of abiotic stress tolerance is more easily accessible. Also the costs of molecular techniques are reduced and in some cases less than phenotyping costs in developed world. However this is very costly in developing world along with requirement of technically advanced protocols. In many situation trait of interest like yield might be poor indicator due to its poor heritability, in such situation a "physiotyping" might be helpful although modest cost to purchase equipments might be required. Thus decision to use yield itself as stress tolerance response or an associated physiological trait or molecular makers depends upon biological along with technological and economic factors. In the following section we will discuss on some of potential traits and their merits and demerits in breeding perspectives.

4. Chlorophyll Associated Traits

4.1. Stay Green

Most of abiotic stress factors can induce the inception of premature senescence (Noodén et al., 1997; Buchanan-Wollaston, 1997; Rosyara et al. 2010a, b; Rosyara et al. 2007). A characteristic consequence of leaf senescence is loss of chlorophyll and progressive decline in photosynthetic ability. Thereby plant defense response to such stress might include any defense mechanism that postpones the onset of senescence and keeps the leaves green till typically active plant production phase such as grain filling completes. Such mechanism to delay senescence is known as stay green which is one of important indicator of tolerance to different types of stresses (Thomas and Smart, 1993). In many crop species, stay-green plants have shown an increased resistance to diseases and drought, are better quality forages for animals, are high in chlorophyll content and are an ideal source of this pigment for food industry, and make pretty ornamental plants over an extended time period (Thomas and Smart 1993; Rosyara et al. 2007).

Stay green has been demonstrated in several economically important plants including *Triticum* (Boyd and Walker, 1972; Rosyara et al., 2007), soybean (Guiamet et al., 1990), rice (Mondal and Choudhuri, 1985; Hidema et al., 1991), corn (Tollenaar and Daynard, 1978; Gentinetta et al., 1986; Crafts-Brandner et al., 1984b), sorghum (Duncan et al., 1981; Gerik and Miller. 1984;), *Festuca* (Thomas, 1987), and *Phaseolus* (Hardwick, 1979). These early studies have shown that genetic variation exists for foliar senescence and has, usually incidentally or empirically, been exploited for crop improvement. Also the studies has led concept of leaf area duration (D) by Watson (1947, 1952), green area duration (G) (Wolfe et al., 1988) or even healthy area duration *(H)* under diseased conditions (Waggoner and Berger, 1987)

Thomas and Smart (1993) has classified senescence related genes in five classes:

(1) Genes controlling the primary metabolic activities of viable cells such as respiratory components, ribosomal RNA synthesis.
(2) Genes that direct installation in the developing mesophyll cell of latent metabolic machinery e.g. vacuolar enzymes, zymogens
(3) Genes that encode growth or carbon assimilation and contributing to progress of senescence by switching off, e.g. nuclear and plastid genes for Calvin cycle enzymes and thylakoid proteins.
(4) Genes particularly turned on at the initiation of senescence
(5) Genes encoding senescence-related activities (e.g. catabolic enzymes) induced *de novo* or increase expression in the process of remobilization.

Thomas and Smart (1993) has classified stay green property into four classes:

- Type A: Delayed initiation of entire senescence syndrome
- Type B: On time start of senescence but proceed at a reduced rate
- Type C: The syndrome could begin and continue on schedule however one or more of the constituent metabolic processes might be hindered.
- Type D: Rapidly killed stay green at harvest e.g. herbarium or frozen food.

Based on this, Thomas and Howarth (2000) have suggested different ways to stay green including knock out of chlorophyll degradation process, inducing perennial tendencies into an annual, use of modified senescence genes, and improving level stay green by marker aided selection analysis followed by QTL analysis. Recent studies have shown association of genes or QTLs with stay green property particular nuclear inheritance (Walulu et al. 1994). Besides exploiting within species diversity, recent advancement in genetic transfer technique provides unlimited opportunity to transfer genes from any one organism to another. In light of this fact, genes associated with *leaf senescence (Sees)* have been cloned from several species and their homologues are isolated (Buchanan-Wollaston, 1997). Around 50 *Sees* have been assigned probable functions in senescence process based on sequence homology.

Another important on going intervention leading to a stay green trait is linked with manipulation of cytokinin status (Thomas and Howarth, 2000). This due to fact that senescence is accompanied by changes in endogenous ethylene, abscisic acid (ABA), and cytokinins, and which is supposed to mediate signaling for the process. In plants transformed with a conditional promoter fused to *ipt* (an *Agrobacterium* gene encoding a limiting step in cytokinin biosynthesis), increase amounts of cytokinins resulting to delayed senescence (Smart et al., 1991). The *ipt* gene encodes isopentenyl transferase (an enzyme that catalyzes the condensation of imethylallylpyrophosphate and 5'-AMP to isopentenyladenosine (iPA) 5'-phosphate). An approach to target the expression of *ipt* to senescing tissues with the promoter from *SAG12* (a senescence-associated gene from *Arabidopsis*) supports the role of cytokinins on plant senescence (Gan and Amasino, 1995). Numerous plants transformed with *SAG12-IPT* have shown significant delays in leaf senescence (Gan and Amasino, 1995; Jordi et al., 2000; Zhang et al., 2000; McCabe et al., 2001; Ori et al. 1999). Zhang et al. (2000) has illustrated for using this process for development of flooding tolerant *Arabidopsis thaliana*. More studies are required to proof applicability of such approach.

4.2. Chlorophyll Content

Leaf chlorophyll content has been important indicator for both biotic as well as abiotic stresses (Reynolds et al., 2001; Shrestha and Rosyara, 2009; Rosyara et al., 2007; Yang et al., 2002; Khavarinejad and Chaparzadeh, 1998; Genc et al., 2010). Leaf chlorophyll measurement has been considered potential trait for heat stress (Rosyara et al. 2007, 2010a,b), cold tolerance (Hund et al. 2005), drought stress (Songsri et al. 2009), light intensity, salinity (Qasim et al., 2003, Khavarinejad and Chaparzadeh, 1998), and nutrient stress (Erley et al. 2007) . In addition to chlorophyll concentration, chlorophyll a:b ratio was used as indicator for heat stress tolerance (Camejo et al., 2005). Increasing salinity decreases content of chlorophyll and the net photosynthetic rate particularly in salt-sensitivity plants (Khavarinejad and Chaparzadeh, 1998). The genes, *ERA1, PP2C, AAPK, PKS3* response to salinity tolerance measure as Chlorophyll content in rice and wheat (Thomson et al 2010, Ma et al. 2007, Genc et al. 2010).

One of important aspect of efficient use of any physiological trait is availability of equipment that can quickly and precisely measure with low error rates nondestructively at individual plant level in short window of time. Due to usefulness of chlorophyll content measurements, efforts to develop perfect chlorophyll meters are under-way. Different types of chlorophyll meter (for example, Minolta Chlorophyll meter Model SPAD 502) particularly of transmittance / absorbance based has been used to make non destructive measurements of leaf greenness or indicator for chlorophyll content (Murdock et al, 1997; Piekielek et al., 1997; Peterson et al., 1993; Rosyara et al., 2010a, b). Although such meters can reduce laboratory efforts using destructive sampling techniques, are not as fast compared to spectral reflectance or fluorescence imaging techniques, will be discussed in following sections. Reflectance based (eg. Field Scout CM 1000) are easier to use that conventional transmittance / absorbance based (SPAD chlorophyll meters) methods. However, reflectance based methods are more affected by sunlight intensity and angle of the sun (Murdock et al., 2004).

4.3. Chlorophyll Fluorescence

Chlorophyll fluorescence has important property of interest in light of understanding level of stress in plants. Photon energy absorbed by photosynthetic pigments drives primary photochemical reactions. General outline for fate of light in leaves and chlorophyll fluorescence is illustrated in figure (5). Energy conversion normally takes place with a high efficiency exceeding 90 % of absorbed quanta (Schreiber et al., 2000). Irradiation excites chlorophyll molecule (*Chl*) molecules to a first excited singlet state stable for less than 10^{-8} s (Holzwarth, 1991) and charge separation at the reaction centre (RC) takes place within picoseconds (Cogdell, 1983). If charge separation does not occur, excited pigment molecules return to ground level and absorbed energy is released as heat (i.e. radiation-less deactivation) and/or Chlorophyll fluorescence (Krause and Weis, 1991). The degree chlorophylls fluorescence is associated with photochemical efficiency, photosynthesis, degree of stress response and stress tolerance. The fundamental definition of terms of chlorophyll fluorescence terminology is illustrated in figure in the box (1).

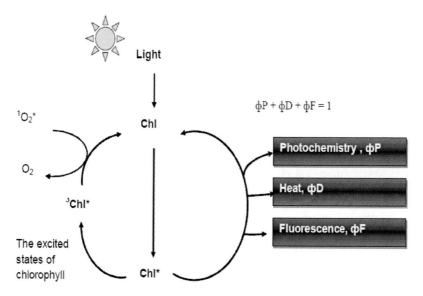

Figure 5. Outline of reaction in photosystem showing evolution of Chlorophyll florescence.

From many years, Chlorophyll fluorescence has been routinely used to monitor the photosynthetic performance of plants both under stressed and non stressed conditions (Yamada 1996; Rosyara et al. 2010a, b; Baker and Rosenqvist, 2004, Rosenqvist, 2001; Ort and Baker, 2002; Leegoood and Edwards, 1996; Allen and Ort, 2001; Ort, 2002). Different chlorophyll fluorescence parameters are in use and will be discussed in following sections.

Box 1. Fluorescence Measurement Parameters

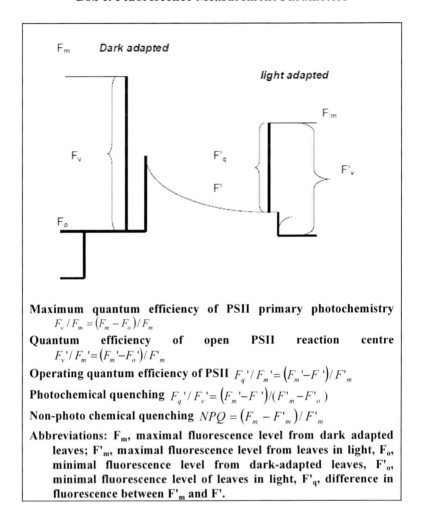

Maximum quantum efficiency of PSII primary photochemistry
$F_v / F_m = (F_m - F_o)/F_m$
Quantum efficiency of open PSII reaction centre
$F_v'/F_m' = (F_m' - F_o')/F_m'$
Operating quantum efficiency of PSII $F_q'/F_m' = (F_m' - F')/F_m'$
Photochemical quenching $F_q'/F_v' = (F_m' - F')/(F_m' - F_o')$
Non-photo chemical quenching $NPQ = (F_m - F_m')/F_m'$
Abbreviations: F_m, maximal fluorescence level from dark adapted leaves; F_m', maximal fluorescence level from leaves in light, F_o, minimal fluorescence level from dark-adapted leaves, F_o', minimal fluorescence level of leaves in light, F_q', difference in fluorescence between F_m' and F'.

F_q'/F_m' measures directly the efficiency of light use in electron transport by PSII. One of the major factors determining such efficiency is the ability of the leaf to remove electrons from the quinone acceptors of PSII. This is directly related to the rate at which the products of photosynthetic electron transport (i.e. NADPH and ATP) are consumed. PSII operating efficiency (estimated by F_q'/F_m'; and CO_2 assimilation) moderate to strong relationship under environmental stresses. The rate of leaf CO_2 assimilation is sensitive to a wide range of environmental stresses including high or low temperatures, high or low light, water deficits or excess, and nutrient deficiency or toxicity(Baker,

1996; Baker and Rosenqvist, 2004). Ability to maintain the PSII quinone acceptors partially oxidized is a key factor in tolerating environmental stresses (Rosenqvist, 2001; Ort and Baker, 2002) and this can be screened for by monitoring F'_q/F'_v. Such relationship allows this trait for gauging tolerance to environmental stresses.

Using chlorophyll fluorescence parameters to measure drought tolerance response should be done with caution (Leegood and Edwards, 1996; Lawlor and Cornic, 2002). For example change in relative water content initially change the stomatal closure, which can result in decreased photosynthesis but do not impact on F_v/F_m however on F'_q/F'_m. In C_3 plants, an increase in photorespiration under the stress conditions might sustain rates of electron transport similar to those observed in non-stressed leaves despite the rate CO_2 assimilation decreasing (Leegood and Edwards, 1996). However, strong stress can result a steady decline of photosynthetic potential, thereby sufficiently large to overcome stomatal limitation (Lawlor and Cornic, 2002). In addition to drought condition, Salinity appears not to influence PSII primary photochemistry at the start (Shabala et al., 1998); however, at high level of salinity reduced leaf water potential causing reduced stomatal conductance reflected in change in F'_q/F'_m. As with drought, fluorescence induction characteristics are changed by salinity, and have the potential to be used for screening for salt-tolerant varieties (Smillie and Nott, 1982).

A most important effect of reductions in temperature is the inhibition of photosynthetic carbon metabolism (Ort, 2002; Leegoood and Edwards, 1996; Allen and Ort, 2001) resulting a decline in the sink for the products of electron transport (ATP and NADPH) and F'_q/F'_m (Andrews et al., 1995; Bruggermann and Linger, 1994; Gray et al., 1997; Fracheboud and Leipner, 2003). Generally, such decline in the rate of utilization of photoreductants and ATP, consequences in a reduction in the PSII efficiency (F'_q/F'_v), accompanied by decreases in F'_v/F'_m (Groom and Baker, 1992; Bruggermann and Linger, 1994; Andrews et al., 1995; Gray et al., 1997; Fracheboud and Leipner, 2003). Inactivation of PSII and thylakoid disorganization are supposed to be key features of high temperature stress and followed by sharp rise in F_o indicating temperature for PSII inactivation (Smillie and Nott, 1979; Havaux, 1993). Both the rise in F_o and a decrease in Fv/ Fm or one of them have been used to determine differences in the response of wheat (Rosyara et al. 2010a,b), potato cultivars (Havaux, 1995) and species of birch (Ranney and Peet, 1994) to high temperatures.

Among different nutrients, Nitrogen shows strong correlation with CO_2 assimilation at high irradiance (Evans, 1989), suggesting potential role of chlorophyll fluorescence as indicator of nitrogen status or stress situation. Results have shown that decreasing nitrogen content of apple leaves decrease F'_q/F'_m; F'_q/F'_v; and F'_v/F'_m (Cheng et al., 2000), although nitrogen contents need to be decline to a very low levels for F_v/F_m to be affected. However, it is more likely that changes in the status other nutrients have little effect on fluorescence property. For example, depletion of sulphur levels in leaves of sugar beet had to reach starvation levels before any further changes in F'_q/F'_m; F'_q/F'_v; and F'_v/F'_m to be observed (Kastori et al., 2000).

Key progresses in the instrumentation for measuring chlorophyll fluorescence non destructively from intact plants along with improved understanding of fluorescence physiology and relationship with tolerance has led widespread use in plant physiological studies. The measurement holds great promise in breeding programs. There are some specific requirements for deployment in a breeding program. One of them is related to speed of measurement to fit into high throughput. Using conventional laboratory technique or handheld chlorophyll fluorometer may not be suitable for breeding experiments due to time and effort require for measurement. More recent advances in the chlorophyll fluorescence imaging has potential role in increasing both the sensitivity and throughput of plant phenomics. Significant progress has been made in instrumentation and technology, for example high throughput screening by growing in 96-well microtitre plates and applying rapid assessment procedures (Berg et al., 1999; Evans, 1999). With development of florescence image system, we can image areas in excess of 80 cm^2 means that we can phenotype many plants simultaneously (Barbagallo et al., 2003; Oxborough, 2004). Method applied includes imaging of fluorescence signals by means of charge coupled device (CCD) cameras (Oxborough, 2004). One of the key advantages of fluorescence imaging is that measurement of florescence is not restricted to small portion of leaves rather whole leaves or large numbers of plants. Results from such imaging of leaves have revealed extremely heterogeneity among leaves and plants (Bro et al., 1996; Genty and Meyer, 1995; Oxborough and Baker, 1997; Siebke and Weis, 1995a, b; Scholes and Rolfe, 1996; Leipner et al., 2001; Meng et al., 2001; Eckstein et al., 1996).

4.4. Spectral Reflectance and Imaging

Solar radiation occurs primarily in the spectral range of the visible and near infrared (NIR) regions in the wavelengths ranging between 400 nm to

2500 nm (Figure 6). In the visible spectrum (400 nm to 750 nm), reflectance by plant leaves or canopies is predominantly low (Figure 6). This is result of absorption by leaf pigments, particularly chlorophyll. With the shift from the visible to the NIR wavelengths, there is a sharp increase in reflectance, is also known as 'red edge'. In the NIR, between 800 nm and 1300 nm, a large amount of the incident radiation is reflected by leaves due to scattering within the leaf mesophyll (Knipling, 1970). Thus under a field situation, greenness and biomass accumulation can be estimated using reflectance measurements in the visible and NIR range (Harris et al., 2007; Penuelas and Filella, 1998; Sims and Gamon, 2002; Carter and Knapp, 2001; Campos et al., 2004). As plant under stress has lower chlorophyll content, this can affect the spectral reflectance significantly.

SPAD readings and the Normalized Difference Vegetation Index (NDVI) are perhaps the most widely method to assess biomass, chlorophyll content and stress condition (Richardson et al., 2002; Markwell et al., 1995; Raun et al., 2001; Penuelas et al., 1997b; Rodriguez et al., 2005; Perry and Roberts, 2008).

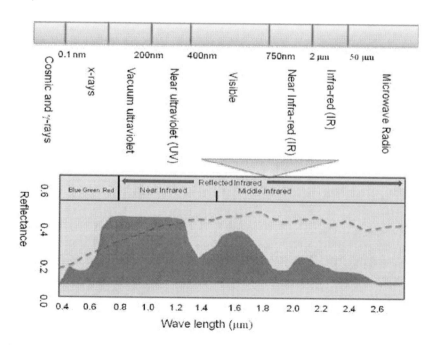

Figure 6. Fundamentals of range of different wave lengths and spectrum spectral reflectance by green plants (shown in solid line) and non green objects like soil (dotted line).

Most important fact is that SPAD and NDVI spectrometers are commercially available. One of important feature of spectral reflectance measurement is that spectral reflectance measure general stress level in plant community, i.e. several factors are confounded (Rodriguez et al., 2005). Even if these methods allow the identification of plants with an overall tolerance to a particular stress, they are unlikely to help in dissecting overall tolerance into genetically tractable traits. Hence, the question comes up as to whether spatial resolution using such color and NIR cameras might be more effective than conventional approaches. Attempts to replace spectral vegetation indices with color image-derived indices are promising and these have been used to estimate traits, such as grain yield under drought conditions (Casadesus et al., 2007), the rate of senescence of wheat (Adamsen et al., 1999), and the early identification of tomato wilting symptoms (Takayama et al., 2009). In addition, such imaging has the a benefit that it can resolve heterogeneity taking place at the leaf, plant, or canopy level (Lenk et al., 2007; Chaerle et al., 2009; Chaerle and Van der Straeten, 2001) and this also makes it possible to separate plants from background soil, particularly important in early and late stages of growth when plants are not able to cover the whole ground.

5. Precision High Throughput Phenotyping

Conventional methods for phenotyping plants are often painstaking and destructive requiring removal of plant biomass for analysis. This precludes for measurement of other traits that require biomass at certain growth stage or need grain harvesting. However, latest developments in high-throughput, non-destructive imaging technologies permit a researcher or breeders to obtain multiple images of the same plant at different time points, growth stages and at different wave lengths, there by presenting novel non-destructive methods for recording quantitative data about plant growth and development under normal and stressed situations (Berger et al. 2010). The technology has already put into place to quantify traits related to drought, salt and heat tolerance in number of crop plants (Berger et al., 2010; Rajendran et al., 2009; Sirault et al., 2009).

Analysis of images and storage of large amount of data is very important in high throughput phenotyping. Such image analysis can be done either developing new software or taking advantage of existing software, such as

MatLab(R) (MathWorks Inc.) or the free-ware package ImageJ (Abramoff et al., 2004). But still challenges are to find measurement parameters that are acceptable to wide range of scientific community. The International Plant Phenomics Network (www.plantphenomics.com) or any other international society might help to establishing such agreement. Recently, a number of high-throughput phenotyping methods have been appearing online to help increase the speed and precision of phenotyping, for example rapid elemental analysis of plant tissue (Baxter et al. 2007) and growth facilities equipped with robotic convey or belts that deliver plants to automatic imaging, watering and weighing stations, for example, ThePlantAccelerator1 in Australia, Crop Design in Belgium and the Leibniz Institute of Plant Genetics and Crop Plant Research in Germany.

Conclusion

Chlorophyll associated traits can play a pivotal role in development of stress tolerant genotypes. Both genomics and phenomics of such traits under stressed situations can help in turning art into science of breeding. For plant breeding purpose there is need to increase efficiency, precision as well as robustness. Developments of such techniques are underway.

References

Aarti, PD; Tanaka, R; Tanaka, A(2006) Effects of oxidative stress on chlorophyll biosynthesis in cucumber (*Cucumis sativus*) cotyledons, *Physiologia Plantarum* 128: 186–197.

Abramoff, MD; Magelhaes, PJ; Ram, SJ (2004) Image processing with Image J. *Biophotonics International* 11: 36–42.

Adamsen, FJ; Pinter, PJ; Barnes, EM; LaMorte, RL; Wall, GW; Leavitt, SW; Kimball, BA (1999) Measuring wheat senescence with a digital camera. *Crop Science* 39:719–724.

Allen, DJ; Ort, DR (2001) Impacts of chilling temperatures on photosynthesis in warm-climate plants. *Trends in Plant Science* 6:36–42.

Amir-Shapira, D; Goldschmidt, EE; Altman, A. (1987). Chlorophyll catabolism in senescing plant tissues: In vivo breakdown intermediates suggest different degradative pathways for citrus fruit and parsley leaves. *Proc. Natl. Acad. Sci. USA* 84: 1901–1905.

Andaya, V; Tai, T (2006) Fine mapping of the qCTS12 locus, a major QTL for seedling cold tolerance in rice, *Theoretical and Applied Genetics* 113:467-475.

Andersson, B; Barber, J (1996). Mechanisms of photodamage and protein degradation during photoinhibition of photosystem II. In Baker, NR. *Photosynthesis and the Environment*, , ed (Dordrecht, The Netherlands: Kluwer Academic Publishers), pp. 101–121.

Andrews, JR; Fryer, MJ; Baker, NR (1995) Characterization of chilling effects on photosynthetic performance of maize crops during early season growth using chlorophyll fluorescence. *Journal of Experimental Botany* 46: 1195–1203.

Baga, M; Chodaparambil, S; Limin, A; Pecar, M; Fowler, D; Chibbar, R (2007) Identification of quantitative trait loci and associated candidate genes for low-temperature tolerance in cold-hardy winter wheat. *Functional and Integrative Genomics* 7:53-68.

Baker NR, Rosenqvist E (2004) Applications of chlorophyll fluorescence can improve crop production strategies: an examination of future possibilities, *Journal of Experimental Botany* 1:15, DOI: 10.1093/jxb/erh196.

Baker, NR (1996) *Photosynthesis and the environment*. Dordrecht: Kluwer Academic Press

Balint, AF; Roder, MS; Hell, R; Galiba, G; Borner, A (2007) Mapping of QTLs affecting copper tolerance and the Cu, Fe, Mn and Zn contents in the shoots of wheat seedlings. *Biologia Plantarum* 51:129-134.

Balint, AF; Szira, F; Roder, MS; Galiba, G; Borner, A (2009) Mapping of loci affecting copper tolerance in wheat—the possible impact of the vernalization gene *Vrn-A1*. *Environmental and Experimental Botany* 65:369-375.

Baxter, I; Ouzzani, M; Orcun, S; Kennedy, B; Jandhyala, SS; Salt DE (2007) Purdue Ionomics Information Management System. An integrated functional genomics platform. *Plant Physiology* 143:600-611.

Berg, D; Tiejten, K; Wollweber, D; Hain, R (1999) From genes to targets: impact of genomics on herbicide discovery. *Proceedings of the BCPC Conference – Weeds* 1999 2, 491–500.

Berger, B; Parent, B; Tester, M (2010) High-throughput shoot imaging to study drought responses. *Journal of Experimental Botany* 61:3519-3528.

Bowler, C; Van Camp, W; Van Montagu, M; Inze, D (1994) Superoxide dismutase in plants. *Crit. Rev. Plant Science* 13: 199-218.

Boyd, WJR; Walker, MG (1972) Variation in chlorophyll a content and stability in wheat flag leaves. *Annals of Botany* 36:87-92.

Boyer, JS (1982) Plant productivity and environments. Science 218:443-447.

Bro, E; Meyer, S; Genty, B (1996) Heterogeneity of leaf CO_2 assimilation during photosynthetic induction. *Plant, Cell and Environment* 19:1349–1358.

Bruggemann, W; Linger, P (1994) Long-term chilling of young tomato plants under low light. IV. Differential responses of chlorophyll fluorescence quenching coefficients in *Lycopersicon* species of different chilling sensitivity. *Plant and Cell Physiology* 35: 585–591.

Buchanan-Wollaston, V (1997) The molecular biology of leaf senescence. *Journal of Experimental Botany* 48:181-199.

Cai, S; Bai, G-H; Zhang, D (2008) Quantitative trait loci for aluminum resistance in Chinese wheat landrace FSW. *Theoretical and Applied Genetics* 117:49-56.

Camejo, D; Rodrıguez, P; Morales, MA; Dellamico, JM; Torrecillas, A; Alarcon, JJ(2005) High temperature effects on photosynthetic activity of two tomato cultivars with different heat susceptibility. *Journal of Plant Physiology* 162: 281–289.

Campos, H; Cooper, A; Habben, JE; Edmeades, GO; Schussler, JR (2004) Improving drought tolerance in maize: a view from industry. *Field Crops Research* 90:19–34.

Capell, T; Escobar, C; Liu, H; Burtin, D; Lepri, O; Christou, P (1998). Overexpression of the oat arginine decarboxylase cDNA in transgenic rice (*Oryza sativa* L.) affects normal development patterns *in vitro* and results in putrescine accumulation in transgenic plants. *Theoretical and Applied Genetics* 97: 246-254.

Carter, GA; Knapp, AK (2001) Leaf optical properties in higher plants: linking spectral characteristics to stress and chlorophyll concentration. *American Journal of Botany* 88: 677–684.

Casadesus, J; Kaya, Y; Bort, J; Nachit, MM; Araus, JL; Amor, S; Ferrazzano, G; Maalouf, F; Maccaferri, M; Martos, V; Ouabbou, H; Villegas, D (2007) Using vegetation indices derived from conventional digital cameras as selection criteria for wheat breeding in water-limited environments. *Annals of Applied Biology* 150: 227–236.

Chaerle, L; Lenk, S; Leinonen, I; Jones, HG; Van Der Straeten, D; Buschmann, C (2009) Multi-sensor plant imaging: towards the development of a stress-catalogue. *Biotechnology Journal* 4:1152–1167.

Chaerle, L; Van der Straeten, D (2001) Seeing is believing: imaging techniques to monitor plant health. *Biochimica et Biophysica Acta* 1519:153–166.

Chappel, J (1995) Biochemistry and molecular biology of the isoprenoid biosynthetic pathway in plant *Annu. Rev. Plant Physiol. Plant Mol. Bioi.* 46:521-47.

Chen, G; Krugman, T; Fahima, T; Chen, K; Hu, Y; Roder, M; Nevo, E; Korol, A (2010) Chromosomal regions controlling seedling drought resistance in Israeli wild barley, *Hordeum spontaneum* C. Koch. *Genetic Resources and Crop Evolution* 57:85-99.

Cheng, L; Fuchigami, LH; Breen, PJ (2000) Light absorption and partitioning in relation to nitrogen content in 'Fuji' apple leaves. *Journal of the American Society of Horticultural Science* 125:581–587.

Cogdell, RJ (1983) Photosynthetic reaction centers. – *Annual Review of Plant Physiology* 34: 21-45.

Collins, NC; Tardieu, F; Tuberosa, R (2008) Quantitative trait loci and crop performance under abiotic stress: where do we stand? *Plant Physiology* 147:469-486.

Crafts-Brandner, SJ; Below, FE; Harper, JE; Hageman, RH (1984) Differential senescence of maize hybrids following ear removal. I. Whole plant. *Plant Physiology* 74:360-367.

Cramer, M (2010) Phosphate as a limiting resource: introduction. *Plant and Soil* 334:1-10.

Dat, J; Van Denabeele, S; Varanova, E; Van Montagou, M; Inze, D; Van Breusegem, F (2000). Dual action of the active oxygen species during plant stress responses. *CMLS Cellular Mol. Life Sci.* 57: 779-795.

Duncan, RR; Bockholt, AJ; Miller, FR (1981) Descriptive comparison of senescent and non-senescent sorghum genotypes. *Agronomy Journal* 73:849-853.

Eckstein, J; Beyschlag, W; Mott, KA; Ryel, RJ (1996) Changes in photon flux can induce stomatal patchiness. *Plant, Cell and Environment* 19:1066–1074.

Efrati, A; Eyal, Y; Paran, I (2005) Molecular mapping of the *chlorophyll retainer* (*cl*) mutation in pepper (*Capsicum* spp.) and screening for candidate genes using tomato ESTs homologous to structural genes of the chlorophyll catabolism pathway *Genome* 48:347-351.

Eisenreich, W; Rohdich, F; Bacher, A (2001) Deoxyxylulose phosphate pathway to terpenoids. *Trends Plant Science* 6: 78–84.

Engel, N; Jenny, TA; Mooser, V; Gossauer, A (1991). Chlorophyll catabolism in *Chlorella protothecoides*. Isolation and structure elucidation of a red bilin derivative. *FEBS Letter* 293: 131–133.

Erley, G; Begum, N; Worku, M; Bänziger, M; Horst, WJ (2007) Leaf senescence induced by nitrogen deficiency as indicator of genotypic differences in nitrogen efficiency in tropical maize. *Journal of Plant Nutrition Soil Science* 170:106–114.

Evans, DA (1999) How can technology feed the world safely and sustainably. In: Brooks, GT; Roberts, TR (eds) Pesticide chemistry and bioscience: *the food-environment challenge*. London, UK: Royal Society of Chemistry, 3–24.

Evans, JR (1989) Photosynthesis and nitrogen relationships in leaves of C3 plants. *Oecologia* 78: 9–19.

Falconer, DS; Mackay, TFC (1996) Introduction to quantitative genetics. Fourth Ed. Longman.

Fleury, D; Jefferies, S; Kuchel, H; Langridge, P (2010) Genetic and genomic tools to improve drought tolerance in wheat. *Journal of Experimental Botany* 61:3211-3222.

Flowers, TJ; Yeo, AR (1995). Breeding for salinity resistance in crop plants: where next? *Australian Journal of Plant Physiol.* 22: 875-884.

Foyer, CH; Descourvieres, P; Kunert, KJ (1994) Protection against oxygen radicals: an important defense mechanism studied in transgenic plants. *Plant Cell Environment* 17: 507-523.

Fracheboud, Y; Leipner, J (2003) The application of chlorophyll fluorescence to study light, temperature and drought stress. In: DeEll, JR; Tiovonen, PMA (eds) *Practical applications of chlorophyll fluorescence in plant biology*. Boston: Kluwer Academic Publishers, pp. 125–150.

Gan, S; Amasino, RM (1995) Inhibition of leaf senescence by autoregulated production of cytokinin. *Science* 270: 1986–1988.

Genc, Y; Oldach, K; Verbyla, A; Lott, G; Hassan, M; Tester, M; Wallwork, H; McDonald, G (2010) Sodium exclusion QTL associated with improved seedling growth in bread wheat under salinity stress. *Theoretical and Applied Genetics* 121:877-894.

Gentinetta, E; Ceppi, D; Lepori, C; Perico, G; Motto, M; Salamini, F (1986) A major gene for delayed senescence in maize. Pattern of photosynthates accumulation and inheritance. *Plant Breeding* 97:193-203.

Genty, B; Meyer, S (1995) Quantitative mapping of leaf photosynthesis using chlorophyll fluorescence imaging. *Australian Journal of Plant Physiology* 22: 277–284.

Gerik, TJ; Miller, FR (1984) Photoperiod and temperature effects on tropically and temperately adapted sorghum. *Field Crops Research* 9:29-40.

Goldschmidt, EE (2001) Chlorophyll decomposition in senescing leaves and ripening fruits: Functional and evolutionary perspectives. *Acta Horticulture* 553: 331–335.

Granick, S (1951) Biosynthesis of Chlorophyll and Related Pigments, *Annual Review of Plant Physiology,* 2: 115-144.

Gray, GR; Cauvin, LP; Sarhan, F; Huner, NPA (1997) Cold acclimation and freezing tolerance. A complex interaction of light and temperature. *Plant Physiology* 114: 467–474.

Gray, J; Janick-Bruckner, D; Bruckner, B; Close, PS; Johal, GS (2002). Light dependent death of maize IlsI cells is mediated by mature chloroplasts. *Plant Physiology* 130:18961907.

Groom, QJ; Baker, NR (1992) Analysis of light-induced depressions of photosynthesis in leaves of a wheat crop during the winter. *Plant Physiology* 100:1217–1223.

Guiamet, JJ; Teeri, JA; Nooden, LD (1990) Effects of nuclear and cytoplasmic genes altering chlorophyll loss on gas exchange during monocarpic senescence. *Plant and Cell Physiology* 31:1123-1130.

Hall, AE (1992) Breeding for heat tolerance. *Plant Breeding Reviews* 10:129–168.

Hardwick, RC (1979) Leaf abscission in varieties of *Phmeolus vulgaris* (L.) and *Glycine max* (L.) Merrill - a correlation with propensity to produce adventitious roots. *Journal of Experimental Botany* 30:795-804.

Harris, K; Subudhi, PK; Borrell, A; Jordan, D; Rosenow, D; Nguyen, H; Klein, P; Klein, R; Mullet, J (2007) Sorghum stay green QTL individually reduce post-flowering drought-induced leaf senescence. *Journal of Experimental Botany* 58:327–338.

Havaux, M (1993) Rapid photosynthetic adaptation to heat stress triggered in potato leaves by moderately elevated temperatures. *Plant, Cell and Environment* 16:461–467.

Havaux, M (1995) Temperature sensitivity of the photochemical function of photosynthesis in potato (*Solanum tuberosum*) and a cultivated Andean hybrid (*Solanum x juzepczukii*). *Journal of Plant Physiology* 146, 47–53.

Hidema, J; Makino, A; Mae, T; Ojima, K (1991) Photosynthetic characteristics of rice leaves aged under different irradiances from full expansion through senescence. *Plant Physiology* 97:1287-1293.

Holzwarth, AR (1991) Excited-state kinetics in chlorophyll systems and its relationship to the functional organization of the photosystems. In: Scheer,

H(ed.): *Chlorophylls*. pp. 1125-1151. CRC Press, Boca Raton – Ann Arbor – Boston –London 1991.

Hörtensteiner, S (1999). Chlorophyll breakdown in higher plants and algae. *Cellular and Molecular Life Science* 56: 330–347.

Hortensteiner, S (2004). The loss of green color during chlorophyll degradation-a prerequisite to prevent cell death? *Planta* 219:191-194.

Hörtensteiner, S (2006). Chlorophyll degradation during senescence. *Annual Review of Plant Biology* 57: 55–77.

Hund, A; Frascaroli, E; Leipner, J; Jompuk, C; Stamp, P; Fracheboud, Y (2005) Cold tolerance of the photosynthetic apparatus: pleiotropic relationship between photosynthetic performance and specific leaf area of maize seedlings, *Molecular Breeding* 16: 321–331.

In-Cheol, J; Se-Jun, O; Seo, J; Choi, W; Song, SI; Kim, CH; Kim, YS; Seo, H-S, Choi, YD, Nahm, BH, Kim, J-K (2003) Expression of a bifunctional fusion of the *Escherichia coli* genes for Trehalose-6-Phosphate Synthase and Trehalose-6-Phosphate Phosphatase in transgenic rice plants increases Trehalose accumulation and abiotic stress tolerance without stunting growth. *Plant Physiology* 131: 516–524.

Ingram, J; Bartels, D (1996). The molecular basis of dehydration tolerance in plants. *Annu. Review of Plant Physiology and Plant Molecular Biology* 47:377-403.

Jakob-Wilk, D; Holland, D; Goldschmidt, EE; Riov, J; Eyal, Y(1999). Chlorophyll breakdown by chlorophyllase: Isolation and functional expression of the Chlase I gene from ethylene-treated citrus fruit and its regulation during development. *Plant Journal* 20: 653-661.

Jordi, W; Schapendonk, A; Davelaar, E; Stoopen, GM; Pot, CS; De Visser, R; Van Rhijn, JA; Gan, S; Amasino, RM (2000) Increased cytokinin levels in transgenic P-SAG12-IPT tobacco plants have large direct and indirect effects on leaf senescence, photosynthesis and N partitioning. *Plant Cell Environment* 23: 279–289.

Kastori, R; Plesnicar, M; Arsenijevic-Maksimovic, I; Petrovic, N; Pankovic, D; Sakac, Z (2000) Photosynthesis, chlorophyll fluorescence and water relations in young sugar beet plants as affected by sulfur supply. *Journal of Plant Nutrition* 23:1037–1049.

Khavarinejad, RA; Chaparzadeh, N (1998). The effects of NaCl and $CaCl_2$ on photosynthesis and growth of alfalfa plants. *Photosynthesis* 35:461-466.

Kirigwi, F; Van Ginkel, M; Brown-Guedira, G; Gill, B; Paulsen, G; Fritz, A (2007) Markers associated with a QTL for grain yield in wheat under drought. *Molecular Breeding* 20:401-413. 50.

Knipling, EB (1970) Physical and physiological basis for the reflectance of visible and near IR radiation from vegetation. *Remote Sensing of Environment* 1:155–159.

Krause, GH; Weis, E (1991) Chlorophyll fluorescence and photosynthesis. The basics. – *Annual Review Plant Physiology Plant Molecular Biology* 42: 313-349.

Kuroki, M; Saito, K; Matsuba, S; Yokogami, N; Shimizu, H; Ando, I; Sato, Y (2007) A quantitative trait locus for cold tolerance at the booting stage on rice chromosome 8. *Theoretical and Applied Genetics* 115:593-600.

Lange BM, Ghassemian M (2003) Genome organization in *Arabidopsis thaliana*: a survey for genes involved in isoprenoid and chlorophyll metabolism. *Plant Molecular Biology* 51: 925–948.

Langridge, P; Paltridge, N; Fincher, G (2006) Functional genomics of abiotic stress tolerance in cereals. *Briefings in Functional Genomics and Proteomics* 4:343-354.

Laperche, A; LeGouis, J; Hanocq, E; Brancourt-Hulmel, M (2008) Modelling nitrogen stress with probe genotypes to assess genetic parameters and genetic determinism of winter wheat tolerance to nitrogen constraint. *Euphytica* 161:259-271.

Lawlor, DW; Cornic, G (2002) Photosynthetic carbon assimilation and associated metabolism in relation to water deficits in higher plants. *Plant, Cell and Environment* 25:275–294.

Lee, SY; Ahn, JH; Cha, YS; Yun, DW; Lee, MC; Ko, JC; Lee, KS; Eun, MY (2007) Mapping QTLs related to salinity tolerance of rice at the young seedling stage. *Plant Breeding* 126:43-46.

Leegood, RC; Edwards, GE (1996) Carbon metabolism and photorespiration: temperature dependence in relation to other environmental factors. In: Baker NR (ed.) *Photosynthesis and the environment*. Dordrecht: Kluwer Academic Publishers, 191–221.

Leipner, J; Oxborough, K; Baker, NR (2001) Primary sites of ozone induced perturbations of photosynthesis in leaves: identification and characterization in *Phaseolus vulgaris* using high resolution chlorophyll fluorescence imaging. *Journal of Experimental Botany* 52: 1689–1696.

Lenk, S; Chaerle, L; Pfundel, EE; Langsdorf, G; Hagenbeek, D; Lichtenthaler, HK; Van der Straeten, D; Buschmann, C (2007) Multispectral fluorescence and reflectance imaging at the leaf level and its possible applications. *Journal of Experimental Botany* 58: 807–814.

Lin, HX; Zhu, MZ; Yano, M; Gao, JP; Liang, ZW; Su, WA; Hu, XH; Ren, ZH; Chao, DY (2004) QTLs for Na+ and K+ uptake of the shoots and roots controlling rice salt tolerance. *Theoretical and Applied Genetics* 108:253-260.

Lou Q, Chen L, Sun Z, Xing Y, Li J, Xu X, Mei H, Luo L (2007) A major QTL associated with cold tolerance at seedling stage in rice (*Oryza sativa* L.). *Euphytica* 158:87-94.

Ma, H-X; Bai, G-H; Lu, W-Z (2006) Quantitative trait loci for aluminum resistance in wheat cultivar Chinese Spring. *Plant and Soil* 283:239-249.

Ma, L; Zhou, E; Huo, N; Zhou, R; Wang, G; Jia J (2007) Genetic analysis of salt tolerance in a recombinant inbred population of wheat (*Triticum aestivum* L.). *Euphytica* 153:109-117.

Mano, Y; Takeda, K (1997) Mapping quantitative trait loci for salt tolerance at germination and the seedling stage in barley (*Hordeum vulgare* L.). *Euphytica* 94:263-272.

Markwell, J; Osterman, JC; Mitchell, JL (1995) Calibration of the Minolta SPAD-502 leaf chlorophyll meter. *Photosynthesis Research* 46:467–472.

Mason, R; Mondal, S; Beecher, F; Pacheco, A; Jampala, B; Ibrahim, A; Hays, D (2010) QTL associated with heat susceptibility index in wheat (*Triticum aestivum* L.) under short-term reproductive stage heat stress. *Euphytica* 174:423-436.

Mathews, K; Malosetti, M; Chapman, S; McIntyre, L; Reynolds, M; Shorter, R; van Eeuwijk, F (2008) Multi-environment QTL mixed models for drought stress adaptation in wheat. *Theoretical and Applied Genetics* 117:1077-1091.

Matile, P; Hörtensteiner, S; Thomas, H; Kräutler, B (1996). Chlorophyll breakdown in senescent leaves. *Plant Physiology* 112: 1403–1409.

McCabe, MS; Garratt, LC; Schepers, F; Jordi, WJRM; Stoopen, GM; Davelaar, E; Van Rhijn, JHA; Power, JB; Davey, MR (2001) Effects of P-SAG12-IPT gene expression on development and senescence in transgenic lettuce. *Plant Physiology* 127: 505–512.

Meng, Q; Siebke, K; Lippert, P; Baur B; Mukherjee, U; Weis, E (2001) Sink-source transition in tobacco leaves visualized using chlorophyll fluorescence imaging. *New Phytologist* 151: 585–596.

Mittler, R (2002). Oxidative stress, antioxidants and stress tolerance. Trends in *Plant Science* 7: 405-410.

Mittler, R; Vanderauwera, S; Gollery, M; Van Breusegem, F (2004). Reactive oxygen gene network of plants. *Trends in Plant Science* 9:490-498.

Mohammadi, V; Zali, AA; Bihamta, MR (2008) Mapping QTLs for heat tolerance in wheat. *Journal of Agricultural Science and Technology* 10:261-267.

Mondal, WA; Choudhuri, MA (1985) Comparison of phosphorus mobilization during monocarpic senescence in rice cultivars with sequential and non-sequential leaf senescence. *Physiologia Plantarum* 65:221-227.

Munns, R; Tester, M (2008) Mechanisms of salinity tolerance. *Annual Review of Plant Biology* 59:651-681.

Murdock, L; Call, D; James J (2004) *Comparison and Use of Chlorophyll Meters on Wheat* (Reflectance vs. Transmittance/Absorbance) AGR-181, UK Cooperative Extension Service.

Murdock, L; Jones, S; Bowley, C; Needham, P; James, J; Howe, P (1997) *Using a chlorophyll meter to make nitrogen recommendation on wheat.* University of Kentucky Cooperative Extension Service. University of Kentucky.

Navakode, S; Weidner, A; Varshney, R; Lohwasser, U; Scholz, U; Borner, A (2009) A QTL analysis of aluminium tolerance in barley, using gene-based markers. *Cereal Research Communications* 37:531-540.

Noodén, LD; Guiamet, JJ; John, I (1997) Senescence mechanisms. *Physiological Plant.* 101: 746–753.

Ori, N; Juarez, MT; Jackson, D; Yamaguchi, J; Banowetz, GM; Hake, S (1999). Leaf senescence is delayed in tobacco plants expressing the maize homeobox gene knotted1 under the control of a senescence-activated promoter. *Plant Cell* 11: 1073–1080.

Ort, DR (2002) Chilling-induced limitations on photosynthesis in warm climate plants: contrasting mechanisms. *Environmental Control in Biology* 40:7–18.

Ort, DR; Baker, NR (2002) A photoprotective role for O_2 as an alternative electron sink in photosynthesis? *Current Opinion of Plant Biology* 5:193–198.

Oxborough, K (2004) Using chlorophyll a fluorescence imaging to monitor photosynthetic performance. In: Papageorgiou, G; Govindjee (eds) *Chlorophyll fluorescence: a signature of photosynthesis*. Dordrecht: Kluwer Academic Publishers

Oxborough, K; Baker, NR (1997) Resolving chlorophyll a fluorescence images of photosynthetic efficiency into photochemical and non-photochemical components—calculation of q_P and F'_v/F'_m without measuring F'_o: *Photosynthesis Research* 54:135–142.

Peleg, ZVI; Fahima, T; Krugman, T; Abbo, S; Yakir, DAN; Korol, AB; Saranga, Y (2009) Genomic dissection of drought resistance in durum wheat and wild emmer wheat recombinant inbreed line population. *Plant Cell and Environment* 32:758-779.

Penuelas, J; Filella, I (1998) Visible and near-infrared reflectance techniques for diagnosing plant physiological status. *Trends in Plant Science* 3: 151–156.

Penuelas, J; Pinol, J; Ogaya, R; Filella, I (1997) Estimation of plant water concentration by the reflectance water index WI (R900/R970). *International Journal of Remote Sensing* 18: 2869–2875.

Perry, EM; Roberts DA (2008) Sensitivity of narrow-band and broadband indices for assessing nitrogen availability and water stress in an annual crop. *Agronomy Journal* 100: 1211–1219.

Peterson, TA; Blackmer, JM; Francis, DD; Schepers, JS (1993) Neb Guide: Using chlorophyll meter to improve N management. Lincoln: *Cooperative Extension*, Institute of Agriculture and Natural Resources, University of Nebraska, Lincoln No. G93-1171-A.

Piekielek, W; Lingenfelter, D; Beagle, D; Fox, R (1997) *The early season chlorophyll test for corn*. The Pennsylvania State University, Cooperative Extension College of Agricultural Sciences.

Pmiinska, A; Tanner, G; Anders, I; Roca, M; Hortensteiner, S (2003). Chlorophyll breakdown: Pheophorbide a oxygenase is a Rieske-type iron-sulfur protein, encoded by the accelerated cell death 1 gene. *Proc. Natl. Acad. Sci. USA* 100:15259-15264

Prasad, S; Bagali, P; Hittalmani S, Shashidhar HE (2000) Molecular mapping of quantitative trait loci associated with seedling tolerance to salt stress in rice(*Oryza sativa* L.).*Current Science* 78:162-164.

Prasil, O; Adir, N; Ohad, I (1992). Dynamics of photosystem II: Mechanism of photoinhibition and recovery process. In Barber J (ed) *Topics in Photosynthesis: The Photosystem Structure, Function and Molecular Biology,* Vol. 11, Elsevier, Amsterdam: pp. 295–384.

Pruzinská, A; Tanner, G; Anders, I; Roca, M; Hörtensteiner, S (2003) Chlorophyll breakdown: pheophorbide a oxygenase is a Rieske-type iron-sulfur protein, encoded by the accelerated cell death 1 gene. *Proc. Natl. Acad. Sci. USA*. 100:15259-15264.

Quarrie, S; Pekic Quarrie, S; Radosevic, R; Rancic, D; Kaminska, A; Barnes, J; Leverington, M; Ceoloni, C; Dodig, D (2006) Dissecting a wheat QTL for yield present in arrange of environments: from the QTL to candidate genes. *Journal of Experimental Botany* 57:2627-2637.

Rajendran, K; Tester, M; Roy, SJ (2009) Quantifying the three main components of salinity tolerance in cereals. *Plant, Cell and Environment* 32:237-249.

Raun, WR; Solie, JB; Johnson, GV; Stone, ML; Lukina, EV; Thomason, WE; Schepers, JS (2001) In-season prediction of potential grain yield in winter wheat using canopy reflectance. *Agronomy Journal* 93: 131–138.

Reynolds, MP; Ortiz-Monasterio, JI; McNab, A (2001) *Application of physiology in wheat breeding*. Mexico, D.F.: CIMMYT.

Richardson, AD; Duigan, SP; Berlyn, GP (2002) An evaluation of noninvasive methods to estimate foliar chlorophyll content. *New Phytologist* 153: 185–194.

Rodriguez, D; Sadras, VO; Christensen, LK; Belford, R (2005) Spatial assessment of the physiological status of wheat crops as affected by water and nitrogen supply using infrared thermal imagery. *Australian Journal of Agricultural Research* 56: 983–993.

Rosenqvist, E (2001) Light acclimation maintains the redox state of PSII electron acceptor QA within a narrow range over a broad range of light intensities. *Photosynthesis Research* 70: 299–310.

Rosyara, UR, Subedi, S; Sharma, RC; Duveiller, E (2009) Spot blotch and terminal heat stress tolerance in south Asian spring wheat Genotypes. *Acta Agronomia Hungerica*. 57: 425-436.

Rosyara, UR; Ghimire, AA; Subedi, S; Sharma, RC (2008a)Variation in south Asian wheat germplasm for seedling drought tolerance traits. *Plant Genetic Resources* 6: 88-93.

Rosyara, UR; Vromman, D, Duveiller E (2008b) Canopy temperature depression as an indication of correlative measure of Spot Blotch resistance and heat stress tolerance in spring wheat. *Journal of Plant Pathology* 90: 103-107.

Rosyara, UR; Khadka, K; Subedi, S; Sharma, RC; Duveiller, E(2007a) Heritability of stay green traits and association with spot blotch resistance in three spring wheat populations. *Journal of Genetetics and Breeding* 61: 1-7.

Rosyara, UR; Pant, K; Duveiller, E; Sharma RC (2007b) Variation in chlorophyll content, anatomical traits and agronomic performance of wheat genotypes differing in spot blotch resistance under natural epiphytotic conditions. *Australasian Plant Pathology* 36: 245–251.

Rosyara, UR; Sharma RC; Duveiller E (2006) Variation of canopy temperature depression and chlorophyll content in spring wheat genotypes and association with foliar blight resistance. *Journal of Plant Breeding Group* 1: 45-52.

Rosyara, U.R; Subdedi, S, Sharma, RC; Duveiller, E (2010a)The effect of spot blotch and heat stress in variation of canopy temperature depression, chlorophyll fluorescence and chlorophyll content of hexaploid wheat genotypes. *Euphytica* 174: 377-390.

Rosyara, UR; Subdedi, S; Sharma, RC; Duveiller, E (2010b). Photochemical efficiency and SPAD value as indirect selection criteria for combined selection of Spot Blotch and terminal heat stress in wheat. *Journal of Phytopathology* 158:813–821.

Salekdeh, GH; Reynolds, M; Bennett, J; Boyer, J (2009) Conceptual framework for drought phenotyping during molecular breeding. *Trends in Plant Science* 14:488-496.

Salisbury, FB; Ross, CW (1986) *Plant Physiology*, 3rd Edition, Wadsworth Publishing Company, USA.

Scholes, JD; Rolfe, SA (1996) Photosynthesis in localized regions of oat leaves infected with crown rust (*Puccinia coronata*): quantitative imaging of chlorophyll fluorescence. *Planta* 199: 573–582.

Schreiber, U; Bilger, W; Hormann, H; Neubauer, C (2000) Chlorophyll fluorescence as a diagnostic tool: Basics and some aspects of practical relevance. – In: Raghavendra, A.S. (ed.): *Photosynthesis. Comprehensive Treatise*. pp. 320-336.Cambridge University Press, Cambridge.

Shabala, S.N.; Shabala, SI; Martynenko, AI; Babourina, O; Newman, IA (1998) Salinity effect on bioelectric activity, growth, Na+ accumulation and chlorophyll fluorescence of maize leaves: a comparative survey and prospects for screening. *Australian Journal of Plant Physiology* 25: 609–616.

Shrestha, SM; Rosyara, UR (2009) *Advances o Helminthosporium leaf blight (Spot Blotch and Tan Spot) research,* Tribhuvan University, Institute of Agriculture and Animal Science, Rampur, Chitwan, Nepal.

Siebke, K; Weis, E (1995a) Imaging of chlorophyll a fluorescence in leaves: topography of photosynthetic oscillations in leaves of *Glechoma hederacea*. *Photosynthesis Research* 45:225–237.

Siebke, K; Weis, E (1995b) Assimilation images of leaves of *Glechoma hederacea*: analysis of non-synchronous stomata related oscillations. *Planta* 196:155–165.

Sims, DA; Gamon, JA (2002) Relationships between leaf pigment content and spectral reflectance across a wide range of species, leaf structures and developmental stages. *Remote Sensing of Environment* 81: 337–354.

Sirault, XRR; James, RA, Furbank, RT (2009) A new screening method for osmotic component of salinity tolerance in cereals using infrared thermography. *Functional Plant Biology* 36:970-977.

Smart, CM; Scofield, SR; Bevan, MW; Dyer, TA (1991). Delayed leaf senescence in tobacco plants transformed with *tmr*, a gene for cytokinin production in *Agrobacterium*. *Plant Cell* 3: 647–656.

Smillie, RM; Nott R (1979) Heat injury in leaves of alpine, temperate and tropical plants. *Australian Journal of Plant Physiology* 6:135–141.

Smillie, RM; Nott, R (1982) Salt tolerance in crop plants monitored by chlorophyll fluorescence in vivo. *Plant Physiology* 70:1049–1954.

Songsri, P; Jogloy, S; Holbrook, CC; Kesmala, T; Vorasoot, N; Akkasaeng, C; Patanothai, A (2009) Association of root, specific leaf area and SPAD chlorophyll meter reading to water use efficiency of peanut under different available soil water, *Agricultural water management* 96: 790 – 798.

Takayama K, Nishina H, Yamamoto N, Hatou K, Arima S. 2009. Early diagnosis of water stress in tomato plant by monitoring of projected area with digital still camera. *Journal of Science and High Technology in Agriculture* 21:59–64.

Tanaka, R; Hirashima, M; Satoh, S; Tanaka, A (2003). The Arabidopsis-accelerated cell death gene ACD1 is involved in oxygenation of pheophorbide a: Inhibition of the pheophorbide a oxygenase activity does not lead to the "stay-green" phenotype in *Arabidopsis*. *Plant Cell Physiology* 44:1266-1274.

Tester, M; Langridge, P (2010) Breeding technologies to increase crop production in a changing world. *Science* 327:818-822.

Tewari, AK; Tripathy, BC (1998) Temperature-Stress-Induced Impairment of Chlorophyll Biosynthetic Reactions in Cucumber and Wheat, *Plant Physiology* 117:851-858.

Thomas, H (1987). *Sid:* a Mendelian locus controlling thylakoid membrane disassembly in senescing leaves of *Festuca pratensis*. *Theoretical and Applied Genetics* 73: 551-555.

Thomas, H; Ougham, H; Hortensteiner, S (2001). Recent advances in the cell biology of chlorophyll catabolism. *Adv. Bor. Res.* 35: 1-52.

Thomas, H; Howarth, CJ (2000) Five ways to stay green. *Journal of Experimental Botany* 51: 329-337.

Thomas, H; Smart, CM (1993) Crops that stay-green. *Ann. Appl. Biol.* 123: 193–233.

Thomson, M; de Ocampo, M; Egdane, J; Rahman, M; Sajise, A; Adorada, D; Tumimbang-Raiz, E; Blumwald, E; Seraj, Z; Singh, R et al (2010) Characterizing the Saltol quantitative trait locus for salinity tolerance in rice. *Rice* 3:148-160.

Todorov, DT; Karanov, EN; Smith, AR; Hall, MA (2003) Chlorophyllase activity and chlorophyll content in wild type and *eti 5* mutant of *Arabidopsis thaliana* subjected to low and high temperatures. *Biol. Plant.* 46: 633–636.

Tollenaar, M; Daynard, TB (1978) Leaf senescence in short-season maize hybrids. *Canadian Journal of Plant Science* 58: 869-874.

Tommasini, R; Vogt, E; Fromenteau, M; Hortensteiner, S; Matile, P; Amrhein, N; Martinoia, E. (1998). An ABC transporter of *Arabidopsis thaliana* has both glutathione conjugate and chlorophyll catabolite transport activity. *Plant Journal* 13:773-780.

Tsuchiya, T; Ohta, H; Okawa, K; Iwamatsu, A; Shimada, H; Matsuda, T; Takamiya, K (1999). Cloning of chlorophyllase, the key enzyme in chlorophyll degradation: Finding of a lipase motif and the induction by methyl jasmonate. *Proc. Natl. Acad. Sci. USA* 96:15362-15367.

von Korff, M; Grando, S; DelGreco, A; This, D; Baum, M; Ceccarelli, S (2008) Quantitative trait loci associated with adaptation to Mediterranean dryland conditions in barley. *Theoretical and Applied Genetics* 117:653-669.

Vranova, E; Inze, D; Van Breusegem, F (2002) Signal transduction during oxidative stress. *Journal of Experimental Botany* 53: 1227-1236.

Waggoner, PE; Berger, RD (1987) Defoliation, disease, and growth. *Phytopathology* 77:393-398.

Walulu, RS; Rosenow, DT; Wester, DB; Nguyen, HT (1994) Inheritance of the stay-green trait in sorghum. *Crop Science* 34: 970–972.

Watson DJ (1952) The physiological basis of variation in yield. *Advances in Agronomy* 4:101-145.

Watson, DJ (1947) Comparative physiological studies on the growth of field crops. I. Variation in net assimilation rate and leaf area between species and varieties, and within and between years. *Annals of Botany* 11:41-76.

Wiithrich, KL; Bovet, L; Hunziker, PE; Donnison, IS; Hortensteiner, S (2000). Molecular cloning, functional expression and characterization of RCC reductase involved in chlorophyll catabolism. *Plant Journal* 21:189-198.

Wissuwa, M; Ismail, AM; Yanagihara, S (2006) Effects of zinc deficiency on rice grow than d genetic factors contributing to tolerance. *Plant Physiology* 142:731-741.

Witcombe, JR; Hollington, PA; Howarth, CJ; Reader, S; Steele, KA (2008) Breeding for abiotic stresses for sustainable agriculture. Philosophical Transactions of the Royal Society B: *Biological Sciences* 363:703-716.

Wolfe, DW; Henderson, DW; Hsiao, TC; Alvino, A (1988) Interactive water and nitrogen effects on senescence of maize. I. Leaf area duration, nitrogen distribution, and yield. *Agronomy Journal* 80:859-864.

Yamada, M; Hidaka, T; Fukamachi, H (1996) Heat tolerance in leaves of tropical fruit crops as measured by chlorophyll fluorescence. *Sci. Hortic.* 67: 39–48.

Yang, M; Ding, G; Shi, L; Feng, J; Xu, F; Meng, J (2010) Quantitative trait loci for root morphology in response to low phosphorus stress in *Brassica napus*. *Theoretical and Applied Genetics* 121:181-193.

Yang, D-L; Jing, R-L; Chang, X-P; Li, W (2007) Identification of quantitative trait loci and environmental interactions for accumulation and remobilization of water-soluble carbohydrates in wheat(*Triticum aestivum* L.) stems. *Genetics* 176:571-584.

Yang, J; Sears, RG; Gill, BS; Paulsen, GM (2002) Quantitative and molecular characterization of heat tolerance in hexaploid wheat. *Euphytica* 126:275-282.

Zhang G-Y, Guo, Y; Chen, S-L; Chen, S-Y (1995) RFLP tagging of a salt tolerance gene in rice. *Plant Science* 110:227-234.

Zhang, J; Van Toai, T; Huynh, L; Preiszner, J (2000) Development of flooding tolerant *Arabidopsis thaliana* by autoregulated cytokinin production. *Molecular Breeding* 6: 135–144.

Zhang, S; Yang, B; Feng, C; Chen, R; Luo, J; Cai, W; Liu, F (2006) Expression of the *Grifola frondosa* Trehalose Synthase gene and improvement of drought-tolerance in Sugarcane *(Saccharum officinarum* L.) *Journal of Integrative Plant Biology* 48: 453–459.

Zhang, SZ; Yang, BP; Feng, CL; Tang, HL (2005). Genetic transformation of tobacco with trehalose synthase gene from *Grifola frondosa* enhances the resistance to drought and salt in tobacco. *Journal of Integrative Plant Biology* 47: 579–587.

In: Chlorophyll
Editors: H. Le, et.al.

ISBN: 978-1-61470-974-9
© 2012 Nova Science Publishers, Inc.

Chapter IV

Chlorophyll Fluorescence Emission Spectra in Photosynthetic Organisms

María Gabriela Lagorio[*]

INQUIMAE/ Dpto. de Química Inorgánica, Analítica y Química Física, Facultad de Ciencias Exactas y Naturales. Universidad de Buenos Aires, Ciudad Universitaria. Pabellón II, 1er piso, C1428EHA, Buenos Aires, Argentina

Abstract

The description, analysis and applications of chlorophyll fluorescence emission spectrum in biological organisms are reviewed. At room temperature, in photosynthetic tissues, chlorophyll fluorescence presents a peak in the red (about 680 nm) due to emission from chlorophyll-a linked to photosystem II and in the far-red (about 735 nm) due to contribution of both photosystems (I and II). Chlorophyll fluorescence analysis is a powerful tool for plant physiologists but its interpretation is usually complicated by the presence of light re-absorption processes inside the photosynthetic tissue. A great deal of important conclusions on photochemistry of plants is usually inferred

---

[*] Fax: 54 11 4576 3341, Phone: 54 11 4576 3378 int.106, e-mail: mgl@qi.fcen.uba.ar.

from leaves fluorescence ratios at different emission wavelengths (typically the fluorescence ratio red/far-red). Nevertheless, most of them are deduced from observed spectra distorted by light re-absorption processes and the resultant conclusions are not reliable. A review of the empirical and theoretical approaches to evaluate the chlorophyll emission spectra inside the photosynthetic organism, free from light re-absorption processes, published to date in literature, is detailed and discussed in this chapter. Not only the steady-state spectrum from plant leaves but also from algae and chlorophyll-containing fruits are discussed. Applications in plant physiology and in monitoring plant stress are presented.

Introduction

It is usually accepted that the fluorescence of chlorophyll was first observed by Brewster in 1834 (Brewster, 1834). He detected a brilliant red color when a strong beam of sun light passed through an alcohol extract of laurel leaves. He also observed that the color turned to orange when thicker extracts samples were considered. Eighteen years later, in 1852, Stokes studied the chlorophyll fluorescence spectroscopically for the first time (Stokes 1852). In 1874, Müler observed that plant leaves also displayed red fluorescence from chlorophyll but this emission was much weaker than that observed for dilute chlorophyll solutions (Müler, 1874).

In green plants both chlorophyll-a and chlorophyll-b are present and with the term chlorophylls we usually refer to the sum of both species. However, chlorophyll fluorescence in plants is due to chlorophyll–a emission, since upon light excitation of chlorophyll-b and other accessory pigments like carotenoids, they transfer their absorbed energy to chlorophyll-a molecules.

The chlorophylls spectra in solution have two main absorption bands, these being in the red and blue parts of the visible spectrum (Figure 1). The bands in the red are known as the 'Q bands' and the bands in the blue to near ultraviolet are known as the 'Soret bands'. The absorption bands seen are characteristic of porphyrins.

When absorbing light energy, chlorophyll-b molecules are excited to the second excited singlet state (S_2) by absorption of light of about 453 nm and to the first excited singlet state (S_1) by absorption around 642 nm. Chlorophyll-a, on the other hand, is excited to S_2 by absorption at around 430 nm and to S_1 excited states by absorption around 662 nm (see Figures 1 and 2).

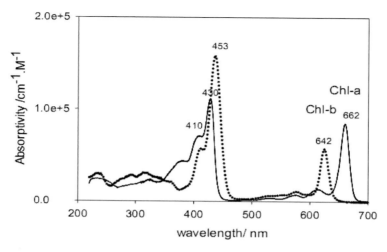

Figure 1. Absorption spectrum for chlorophyll-a (full line) and chlorophyll-b (pointed line) in diethyl ether. Figure drawn from data extracted from "PhotochemCAD. A Computer-Aided Design and Research Tool in Photochemistry and Photobiology," Du, H.; Fuh, R.-C. A.; Li, J.; Corkan, L. A.; Lindsey, J. S. Photochem Photobiol. 1998, 68, 141–142 and "PhotochemCAD 2. A Refined Program with Accompanying Spectral Databases for Photochemical Calculations," Dixon, J. M.; Taniguchi, M.; Lindsey, J. S. Photochem Photobiol. 2005, 81, 212–213.

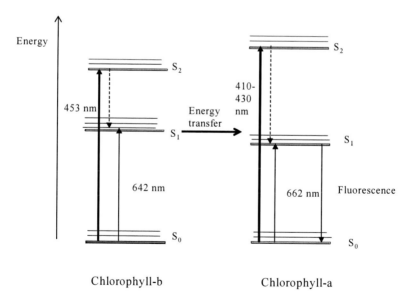

Figure 2. Energy level diagrams of Chlorophyll-b and Chlorophyll-a, indicating energy transfer from the lowest singlet excited state of Chlorophyll-b to that of Chlorophyll-a.

For both types of molecules there is a rapid deactivation from S_2 to S_1 states. Afterwards, energy transfer from the S_1 state of Chlorophyll-b to the S_1 state of Chlorophyll-a takes place (probably by a resonance mechanism) (Rabinowitch and Govindjee, 1969). In this way, Chlorophyll-a becomes the energy acceptor species that subsequently undergoes different deactivation pathways (Figure 2).

The energy transfer between Chlorophyll-b and Chlorophyll-a has a high probability because the process is "downwards". In fact, after the internal conversion (vibrational energy dissipation) from the S_2 to the S_1 state of the donor pigment (Chlorophyll-b), states are reached that are in resonance with certain vibrating states of the acceptor (Chlorophyll-a) giving place to the process of energy transfer. This process is controlled by i) the overlapping between the fluorescence band of the donor and the absorption band of the acceptor that leads to dipole-dipole interaction, ii) the distance between the donor and the acceptor molecules and iii) the orientation factor related to the orientation of the dipoles of the donor and the acceptor molecules (Rabinowitch and Govindjee, 1969).

Phycocyanins, in cyanobacteria, and fucoxanthin, in diatoms and green algae, play the function of energy donors as chlorophyll-b in green plants (Rabinowitch and Govindjee, 1969).

The fluorescence emission spectrum of Chlorophyll-a in an ether solution is shown in Figure 3.

The Q(0,0) band (669 nm) is far more intense than Q(0,1) band (721 nm) following the general rule of emission spectroscopy that fluorescence spectra are mirror images to the first electronic absorption band (Lakowicz, 2006; Parson, 2009).

In photosynthetic organisms, however, Chlorophyll-a is linked to proteins in a complex way and its emission spectrum vary from that shown for solutions as it will be shown in this chapter later.

In photosynthetic tissues, the light energy directly absorbed or received from energy transfer by chlorophyll-a molecules in plant leaves, can undergo different deactivation pathways: i. Initiate the photosynthesis process by beginning the electron transfer process, ii. Dissipate energy excess as heat or iii. Emit energy excess as fluorescence light. All three processes take place in competition meaning that an increase in the efficiency of one of them leads to a decrease in the efficiency of the other two (Maxwell and Johnson, 2000).

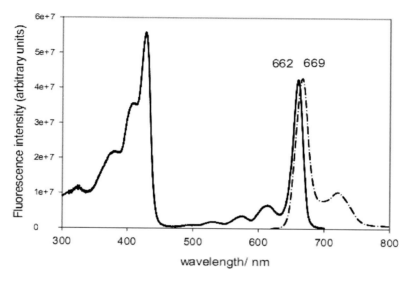

Figure 3. Fluorescence emission (dashed and pointed line) and absorption (full line) spectra of Chlorophyll-a in diethyl ether. Figure drawn from data extracted from "PhotochemCAD. A Computer-Aided Design and Research Tool in Photochemistry and Photobiology," Du, H.; Fuh, R.-C. A.; Li, J.; Corkan, L. A.; Lindsey, J. S. Photochem. Photobiol. 1998, 68, 141–142 and "PhotochemCAD 2. A Refined Program with Accompanying Spectral Databases for Photochemical Calculations," Dixon, J. M.; Taniguchi, M.; Lindsey, J. S. Photochem. Photobiol. 2005, 81, 212–213.

In plants, the major part of the light absorbed by the leaves (more than 80%) is used in the process of photosynthesis (Buschmann, 2007). A small part of the absorbed radiation is dissipated as heat and another small part (less than 2%) (Moya and Cerovic, 2004) is emitted as fluorescence. The chlorophyll fluorescence emitted by leaves is very low compared with the reflected light from them. (Zarco Tejada et al., 2000; Buschman, 2007). The contribution of Chlorophyll emission to the apparent vegetation reflectance is so not very important (Smorenburg et al., 2002; Zarco-Tejada et al., 2000) but it is not completely negligible in the red and far-red regions. In fact, about 10-25% of the apparent reflectance at 685 nm and 2-6% at 740 nm was found to be due to Chlorophyll fluorescence in Corn foliage (Entcheva Campbell et al., 2002).

When excited in the UV region plant leaves also emit blue and green fluorescence (Broglia, 1993; Yaryura et al., 2009). The blue fluorescence is characterized by a peak at about 450 nm and the green fluorescence by a maximum at about 530 nm. Lang et al. (Lang et al., 1991) reported that this emission is mainly originated in the cell-wall and they also observed it was higher for the adaxial (upper) face than for the abaxial (lower) face of leaves. They also stated that phenolic substances such as chlorogenic acid, caffeic acid, coumarins and stilbenes may be responsible for the blue fluorescence emission, whereas substances like berberine and quercetin are responsible for the green fluorescence (Lang et al., 1991). As it does not involve chlorophyll contribution, blue-green emission of leaves will not be described in detail in the present chapter.

At room temperature leaves emit fluorescence in the red (about 685 nm) and in the far-red due to chlorophyll-a. The ratio between the fluorescence emission in the red and in the far red is called in bibliography the "fluorescence ratio". This fluorescence ratio varies when there are changes in the photosynthetic apparatus, stress situations, and nutrient deficiency in plants among others (Buschman, 2007).

Due to the photosynthetic process, the features of chlorophyll-a emission from plants and from related organisms strongly differ from those observed for pigments in inert matrices. These differences are revealed in nature in a fascinating way to show the relationship between the intricate processes of electron transfer and light interaction taking place in the photosynthetic organisms. These characteristics and their interpretations to date will be unveiled in the next sections.

Chlorophyll Fluorescence and Photosynthesis

Photosynthesis may be defined as the physicochemical process by which photosynthetic organisms use light energy to drive the synthesis of organic compounds. In oxygenic photosynthetic organisms (plants, algae and some type of bacteria) water is oxidized to molecular oxygen by the Photosystem II reaction center. In anoxygenic photosynthetic organisms (purple bacteria, green sulfur bacteria, green gliding bacteria and gram positive bacteria) electrons are extracted from molecules other than water (H_2S for instance) and oxygen is not evolved in the process.

In general, photosynthesis reaction may be written as:

$$CO_2 + 2\ H_2A + Light \rightarrow [CH_2O] + 2A + H_2O$$

In oxygenic photosynthesis, H_2A is water and $2A$ is O_2. Oxygenic photosynthetic organisms contain two reaction centers (Photosystem II and Photosystem I) that work concomitantly but in series and they depend on chlorophyll for the conversion of light energy into chemical free energy. Anoxygenic photosynthetic bacteria, on the other hand, have only one type of reaction center which may be either similar to Photosystem II or to Photosystem I, and the conversion of light energy is dependent on bacteriochlorophyll (Whitmarsh and Govindjee, 1999).

The photosystem II (PSII) is made of different protein molecules bound to Chlorophyll-a molecules and other pigments. At the center, a special pair of Chlorophyll-a molecules known as P680 (reaction center) is found (680 refers to the light wavelength of maximum absorption for this species). The oxygen evolving complex is also found at this center. Surrounding the reaction center is the light-harvesting complex (LHC) which contains several protein molecules associated with both Chlorophyll-a and Chlorophyll-b.

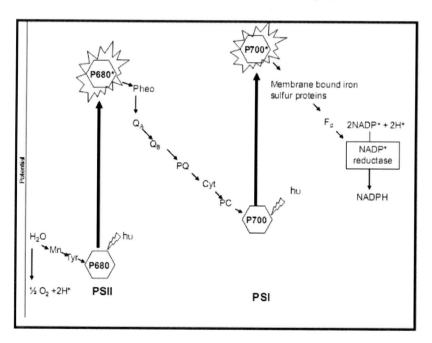

Figure 4. Schematic description of the Photosynthetic process. The abbreviations state for: Mn: Manganese complex bound to PSII, Tyr: a tyrosine in PSII, Pheo: Pheophytin, Q: Quinone, PQ: Plastoquinone, CyT: Cytochrome bf complex, PC: Plastocyanin, Fd: Ferredoxin, P680 and P700 are the reaction centers for PSII and PSI respectively.

The LHC acts as an antenna absorbing photons and transferring their energy to the reaction center. Photosystem I (PSI) is another complex composed of protein molecules bound to Chlorophyll-a and carotenoids. It has also its reaction center formed by a pair of Chlorophyll-a molecules known as P700 (700 refers to the light wavelength of maximum absorption for this species). The photosynthetic apparatus in eukaryotic cells, containing both PSII and PSI, is located in Chloroplasts (Lehninger, 1975). In Figure 4, the photosynthetic process is schematically summarized.

Chlorophyll-a placed at P680 in PSII is excited by the absorbed light or by the energy transferred from the light harvesting complex. Once excited, it can transfer one electron to pheophytin (Pheo) to initiate the photosynthetic process, it can dissipate the excess energy as heat or it can emit fluorescence. As these three processes proceed in competition, fluorescence emission is strongly related to the photosynthetic performance of the organism and to the heat dissipation. In fact, in 1931, Hans Kautsky and A. Hirsch from the Chemisches Institut der Universität-Heidelberg in Germany correlated experimentally Chlorophyll fluorescence with the CO_2 assimilation as a function of time. They worked with dark-adapted leaves that were illuminated for the fluorescence experiments (Kautsky and Hirsch, 1931).

Fluorescence emission from Chlorophyll-a can originate in PSII and/or PSI. An analysis of these emissions will be presented in the next parts of this chapter.

Variable and Non-Variable Chlorophyll Fluorescence

Variable Chlorophyll Fluorescence. Kautsky Kinetics

When transferring photosynthetic material from the dark to the light, an increase in the yield of the chlorophyll fluorescence is observed. This phenomenon was first reported by Kautsky et al. in 1960 (Kautsky et al., 1960). When Chlorophyll-a in PSII is excited, it transfers electrons to the primary acceptors in the photosynthetic chain. Once the quinone Q_A (see Figure 4) has accepted an electron it is not able to accept another until it has been transferred to the next acceptor Q_B. During this time, the reaction center is described as "closed" and the fluorescence emission increases from an initial value F_0 up to a maximum value Fm (Figure 5). This period of time is usually in the order of 1 second. Later, fluorescence starts to fall in a process called "fluorescence quenching" that last several minutes to finally reach a stationary state (Fs). The fluorescence quenching has a photochemical and a non-photochemical contribution. The photochemical quenching (q_p) is due to activation of enzymes involved in the carbon metabolism induced by light and the opening of stomata. The non-photochemical quenching (q_{Np}) is due to an increase in the yields of heat dissipation. (Maxwell and Johnson, 2000).

The described variation in chlorophyll fluorescence is known as Kautsky kinetics. To explore Kautsky kinetics, pulse amplitude modulated (PAM) chlorophyll-fluorometers are used nowadays. In these instruments, a pulse modulated beam induces a pulsed fluorescence signal from the sample where ambient light and non-pulsed fluorescence signal are discarded (Lichtenthaler et al., 2005).

A typical signal obtained for a photosynthetic organism with a PAM fluorometer is shown in Figure 5.

The modulated beam (measuring light) is composed of weak red pulses from a LED (wavelength 650 nm). The pulse duration may vary between 1 to 1.8 µs and the integrated amount of light from the modulated beam is lower than 0.05 $\mu mol.m^{-2}.s^{-1}$.

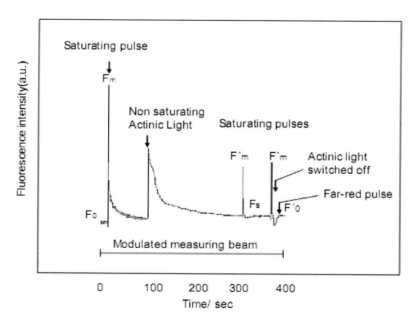

Figure 5. Variable chlorophyll fluorescence recorded with a pulse-modulated fluorometer for a standard plant leaf.

The measuring frequency changes from about 1.6 KHz to 100 KHz for F_0 and F_m measurements respectively. The actinic light is a continuous red radiation with a photon flux in the range 80-300 $\mu mol.m^{-2}.s^{-1}$. It is much more intense than the measuring modulated beam as to induce photosynthesis, but not so high as to saturate the photosynthetic chain. The saturating light is a very intense white light (photon flux ca. 3000-18000 $\mu mol.m^{-2}.s^{-1}$) that saturates the photosynthetic process and closes the reaction centers, so that the whole absorbed energy is dissipated as fluorescence and heat. In most PAM fluorometers, there exists also the possibility to irradiate in the far-red region with a LED (735 nm). This source is applied to excite only PSI in order to remove the electrons from the acceptor part of PSII opening the PSII reaction center and allowing a correct F_0' determination. (Lichtenthaler et al., 2005; Hansatech FMS1 PAM Chlorophyll fluorometer operation manual, 2009)

To derive the photochemical parameters from fluorescence measurements during Kautsky kinetics, the following simplified mechanism may be considered:

Chl-a $\longrightarrow$ Chl-a*

Chl-a* $\longrightarrow$ Photosynthesis $\quad k_p$

Chl-a* $\longrightarrow$ Chl-a + Fluorescence $\quad k_f$

Chl-a* $\longrightarrow$ Chl-a + Heat $\quad k_d$

where k_p is the rate constant for the reaction leading to the photosynthetic process, k_f is the fluorescence constant rate and k_d is the constant rate for heat dissipation and others.

The quantum yield of photosynthesis (Φ_p) may be then calculated as (Govindjee, 2004):

$$\Phi_p = \frac{k_p}{k_p + k_f + k_d} \tag{1}$$

The quantum yield for the minimal Chlorophyll-a fluorescence is expressed by equation (2):

$$\Phi_{f_0} = \frac{k_f}{k_p + k_f + k_d} \qquad (2)$$

and for the maximal chlorophyll-a fluorescence (attained when reaction centers are closed, i.e $k_p \to 0$) we can write:

$$\Phi_{fm} = \frac{k_f}{k_f + k_d} \qquad (3)$$

From equations (1)-(3) it can be deduced that

$$\frac{\Phi_{fm} - \Phi_{f_0}}{\Phi_{fm}} = \frac{k_p}{k_p + k_f + k_d} = \Phi_p \qquad (4)$$

So, $(F_m-F_0)/F_m$ (that is a equivalent to $\dfrac{\Phi_{fm} - \Phi_{f_0}}{\Phi_{fm}}$) is a measure of the maximal photosynthesis quantum yield. No changes either in the absorption cross section of chlorophyll species or in the intensity of the excitation beam are involved as assumptions in these equations (Govindjee, 2004).

As several authors support that most Chlorophyll-a fluorescence at room temperature comes from PSII (Butler W., 1978), the ratio $(F_m-F_0)/F_m$ is usually referred as the maximum quantum yield of PSII photochemistry. However, a contribution of PSI is actually present in the fluorescence emission (Pfundel, 1998). This contribution is discussed later in this chapter.

In bibliography other parameters deduced from chlorophyll-fluorescence measurements have been derived. Details about each parameter are given in the references respectively cited in Table 1.

In Table 1 the photosynthetic parameters obtained with a PAM fluorometer are summarized and briefly described.

Table 1. Photosynthetic parameters derived from the analysis of Kautsky kinetics

Parameter	Term	Description
F_0	Initial fluorescence	Fluorescence yield following dark adaptation when all of the PSII reaction centers and electron acceptor molecules are fully oxidized (Open reaction center) (Lichtenthaler et al., 2005)
F_m	Maximum fluorescence	F_m is attained when a dark-adapted photosynthetic sample is exposed to an intense saturating pulse of light (Lichtenthaler et al., 2005)
F_v/F_m	Ratio between variable and maximum fluorescence	Maximum quantum yield of PSII photochemistry in the dark adapted state, calculated as $(F_m-F_0)/F_m$ (Kitajima and Butler, 1975)
F_v/F_0	Ratio between variable and initial fluorescence	Parameter calculated as $(F_m-F_0)/F_0$ (Lichtenthaler and Buschmann, 1984; Babani and Lichtenthaler, 1996)
F_s	Steady state fluorescence	F_s corresponds to a steady state obtained for light adapted photosynthetic materials after several minutes of irradiation (Lichtenthaler et al.; 2005)
F'_m	Light adapted fluorescence maximum	F'_m is attained when a light-adapted photosynthetic sample is exposed to an intense saturating pulse of light (Lichtenthaler et al., 2005)
Φ_{PSII}	Quantum efficiency of PSII	$(F'_m-F_s)/F'_m$ (Genty et al., 1989)
q_p	Photochemical quenching coefficient	$(F'_m-F_s)/(F'_m-F_0)$ (Bilger and Schreiber, 1986)
q_{Np}	Non-photochemical quenching	$(F_m-F'_m)/(F_m-F_0)$ (Bilger and Schreiber, 1986)
NPQ	Alternative definition of non-photochemical quenching	$(F_m-F'_m)/F'_m$ (Bilger and Björkman, 1990)
Φ_{PSIIR}	Corrected Quantum efficiency of PSII	$(F'_m-F'_0)/F'_m$ The sample is shaded transiently and a far-red light is applied to excite preferentially PSI relative to PSII. It produces "opening" of the reaction centers and it allows measurement of light adapted F_0 (F'_0) (Hansatech, 2009 and Lichtenthaler et al., 2005)

Spectral Distribution of Chlorophyll Fluorescence in Vivo

When irradiating a photosynthetical organism with a low photon flux (usually lower than 20 μmol.m^{-2}.s^{-1}) as not to induce significantly Kautsky kinetics, a spectrum invariable in time may be obtained (Ramos and Lagorio, 2004). Experimentally, constant chlorophyll spectra may be obtained for the initial state (F_0) or for the steady-state situation Fs.

Chlorophyll fluorescence spectrum is characterized by two bands: one in the red (about 680 nm usually denoted as F_{red}) and one in the far-red (about 735 to 740 nm denoted as $F_{far-red}$) (Figure 6).

At room temperature PSII contributes to both F_{red} and $F_{far-red}$ bands wile PSI only to the $F_{far-red}$ (Mazzinghi et al.,1994, Pfundel, 1998).

The fluorescence ratio ($F_{red}/F_{far-red}$) is an important tool that was extensively studied as a function of stress and chlorophyll content in plants. Several authors affirm that the chlorophyll fluorescence ratio varies from different plant species but this is exclusively due to differences in the chlorophyll content (Lang et al., 1991; Hak et al., 1990; Lichtenthaler and Rinderle, 1988, Lichtenthaler et al., 1990).

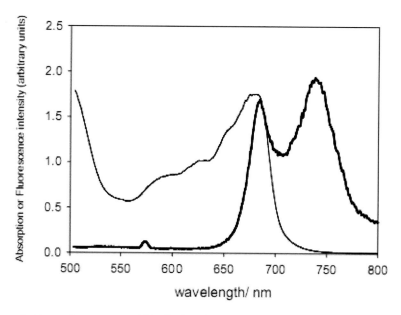

Figure 6. Absorption spectrum (thin line) and F_0 fluorescence emission spectrum corrected by the detector response to wavelengths (thick line) for a leaf of *Ficus benjamina*. Excitation wavelength: 460 nm.

However other authors (Virgin, 1954) opposed to this statement. According Buschmann (Buschmann, 2007), the fluorescence ratio of leaves decreases with increasing chlorophyll content and he states this is a good inverse indicator of the chlorophyll concentration in leaves. Other studies (Agati, 1998; Pfundel, 1998) have shown that the fluorescence ratio at room temperature would depend on both the chlorophyll concentration and photosystems ratio (PSII/PSI). Nevertheless, the fluorescence ratio was used to estimate approximately the Chlorophyll content of leaves (Lichtenthaler et al., 1990).

Mazzinghi et al. (Mazzinghi et al., 1994) state that the fluorescence ratio can give information about the photosynthetical activity of a plant. They have found that this ratio changed during a diurnal cycle. They also affirm that changes in the fluorescence ratio may reflect variations in the activity of the two photosystems since F_{red} is due to PSII emission while $F_{far-red}$ is due to both PSII and PSI emissions. These authors state that inferring chlorophyll concentration from red fluorescence spectra of leaves is delicate as the fluorescence ratio is affected by temperature and by the light intensity impinging onto the leaf. In a detailed work, Agati analyzed changes in Chlorophyll fluorescence under different environmental factors (Agati, 1998) . Working in steady state conditions under low photon flux irradiation, both F_{red} and $F_{far-red}$ increased but the fluorescence ratio decreased when temperature diminished from 25 to 4 °C. This result was interpreted in terms of a lowering in thylakoids membrane fluidity at low temperatures leading to inhibition in the re-oxidation of plastoquinones with a consequent increase in fluorescence emission. The photochemical quenching decreased with decreasing temperature while the non-photochemical quenching was slightly affected. Photochemical and non-photochemical quenching act differently on PSII and PSI and this would be the reason why the fluorescence ratio changes with leaf temperature (Agati, 1998).

The experimental fluorescence spectra of leaves change their spectral distribution for different excitation wavelengths (Agati, 1998; Ramos and Lagorio 2004).

At high photon flux (2000 $mol.m^{-2}.s^{-1}$), keeping the temperature constant, F_{red}, $F_{far-red}$ and the fluorescence ratio decreased during the irradiation time. This decrease is first rapid due to the Kautsky kinetics and then slow due to the activation of photoinhibition-protecting mechanisms. It shows that high irradiation fluxes induce changes in the spectral distribution of chlorophyll fluorescence in vivo, before reaching the steady state. In fact, the fluorescence ratio $F_{red}/F_{far-red}$ increases on going from F_0 to F_m during the Kautsky kinetics.

At F_s the fluorescence ratio is again similar to the value at F_0. (Franck et al,. 2005)

Contribution of PSII and PSI Fluorescence Emission at Room Temperature

There are several manuscripts that state that chlorophyll fluorescence is mainly due to PSII emission at room temperature. They affirm that only a small contribution of PSI is present under this condition (Krause and Weis, 1991; Govindjee, 1995; Buschman, 2007). However, several authors demonstrated that the effects of PSI fluorescence are significant at room temperature and showed that they should be taken into account when analyzing chlorophyll fluorescence (Pfündel, 1998; Agati et al., 2000, Franck et al. 2002).

During the evolution from open to closed reaction centers, PSII fluorescence increases several times while PSI emission is independent of the state of its reaction center (Butler, 1978; Briantais et al., 1986; Pfündel, 1998). It is the reason why the "variable" chlorophyll fluorescence in vivo, measured at room temperature, is correctly attributed in literature only to PSII (Pfündel, 1998). Nevertheless, there is a constant contribution to fluorescence provided by PSI that should be taken in mind.

Under irradiation with low light intensity (low enough as not to induce variable fluorescence), i.e. when recording F_0, PSI contribution to emission at wavelengths longer than 700 nm is very important and may vary between 30% (for C3 plants) and 50% (for C4 species) about (Pfündel, 1998). Under irradiation with high light intensity (induction of variable fluorescence taking place), PSI contribution to emission at wavelengths longer than 700 nm is less important. For the F_m state this contribution can vary from 6 to 12% approximately. (Pfündel, 1998; Peterson et al., 2001). These results show that PSI relative contribution is very important when working at room temperature under irradiation with low photon flux (F_0).

For high irradiation intensities the relative contribution is much lower. Even though, Pfundel has shown that the constant contribution to fluorescence provided by PSI at room temperature can distort the quantitative analysis from which the PSII yield is calculated (Kautsky kinetics) (Pfundel, 1998). So, the fact that variable chlorophyll fluorescence is only due to PSII does not imply that PSI contribution has no effect in the ulterior calculations performed to obtain the photosynthetical parameters.

Pfündel found a linear correlation between F_v/F_m and the ratio $F_{red}/F_{far-red}$ (more specifically F_{735}/F_{685}) at 77K and he concluded that F_v/F_m ratios are usually underestimated due to the contribution of PSI that is not taken into account.

Considering PSI contribution, the parameter Fv/Fm introduced in Table 1 should be then strictly re-written as:

$$\frac{F_v}{F_m} = \frac{F_m^{PSII} + F_m^{PSI} - (F_0^{PSII} + F_0^{PSI})}{F_m^{PSII} + F_m^{PSI}} = \frac{F_m^{PSII} - F_0^{PSII}}{F_m^{PSII} + F_m^{PSI}} \quad (5)$$

where F_m^{PSI} and F_0^{PSI} were cancelled to account for the fact that there is no variable fluorescence for PSI.

Equation (5) may be re-written so that:

$$\frac{F_m}{F_v} = \frac{1}{\Phi_V^{PSII}} (1 + \frac{F_m^{PSI}}{F_m^{PSII}}) \quad (6)$$

where Φ_V^{PSII} denotes the Fv/Fm of dark adapted PSII. (Pfundel, 1998).

Equation (6) shows that including explicitly PSI contribution and due to the factor $(1+\frac{F_m^{PSI}}{F_m^{PSII}})$, it is not strictly correct to attribute Fv/Fm to the maximum photochemical efficiency of PSII.

Several photosynthetic parameters derived from the measurement of the variable fluorescence as a function of time (Kautsky kinetics) may be affected by PSI contribution. After analyzing the corresponding equations, it is possible to conclude that the involvement of PSI can distort the PSII quantum yield calculated according to Genty et al. (Genty et al., 1989) and the non-photochemical quenching coefficient NPQ (Bilger and Björkman, 1990). On the contrary, PSI contribution should not affect the coefficients for photochemical and non-photochemical fluorescence quenching as defined by Bilger and Schreiber (Bilger and Schreiber et al, 1986) because this effect should cancel in the calculations (see Table 1).

As a general conclusion, PS I fluorescence cannot be overlooked whenever a detailed analysis on steady state fluorescence or on fluorescence kinetics from a photosynthetic tissue at room temperature is performed. Additionally, it should be taken in mind that the magnitude of the irradiation intensity is extremely significant to assess the relative contribution of PSI emission and it is particularly important when analyzing F_0 spectra.

Spectral Distribution of Chlorophyll Fluorescence in Chloroplasts. Corrections for Light Re-Absorption Processes

The observed chlorophyll fluorescence from intact leaves is usually distorted by light re-absorption processes inside the foliar tissue. As a consequence, the observed spectral distribution differs from that emerging from the chloroplasts. As we have explained before, information about the plant physiological state and the photosystem ratio can be derived from the fluorescence ratio. However, the observed ratio is affected additionally by artifacts due to the phenomenon of light re-absorption. Efforts have been made in literature to retrieve corrected spectra from the observed ones.

Agati et al. developed a theoretical model to correct experimental fluorescence spectra from leaves assuming an infinitesimal layer in the leaf where the excitation beam is attenuated exponentially (Agati et al., 1993). In this approach, a fluorescence beam is generated at the infinitesimal layer, it is exponentially attenuated by light re-absorption and it is emitted isotropically inside the leaf. The light scattering is taken into account through considering an effective path length and lateral light losses are neglected (low leaf thickness), The validation of the method is performed by comparing emission spectra corrected by re-absorption for both aurea mutant and wild type of tomato leaves, which differ in chlorophyll concentration and chloroplast ultrastructure.

Gitelson et al. developed an empirical correction model that proposes a correction factor which depends on the leaf reflectance and transmittance (Gitelson et al., 1998). One of the feeble points of this method is the fact that their correction factor does not depend on the excitation wavelength.

Later, Ramos and Lagorio applied a correction model (Ramos and Lagorio, 2004), developed previously for fluorescent dyes immobilized on inert materials (Lagorio et al., 1998; Lagorio et al., 2001; Iriel et al., 2002; Rodriguez et al., 2004). The approach consists of a two-flux model that assumes that the leaves behave as an ideal diffuser. The emission produced in each volume element is decomposed into two-photon flows having the same magnitude but opposite direction. They validated the application of the model to leaves by matching the spectra corrected by light re-.absorption processes with those recorded for a thin layer of isolated chloroplasts of the same plant species deposited on a quartz plate. It was proved for different excitation wavelengths. In figures 7, 8 and 9 we can see the results and the good agreement found.

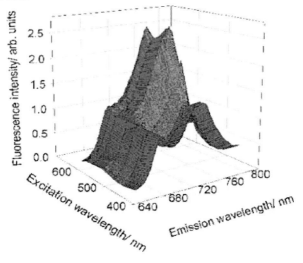

Figure 7. Experimental emission spectra of *Ficus benjamina* leaves at different excitation wavelengths, normalized at 685 nm from (Ramos and Lagorio, 2004 – Reproduced by permission of the Royal Society of Chemistry (RSC) for the European Society for Photobiology, the European Photochemistry Association, and the RSC).

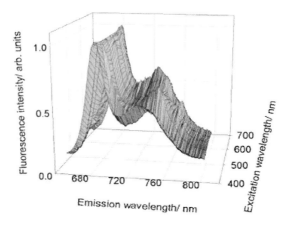

Figure 8. Emission spectra of *Ficus benjamina* leaves corrected by light reabsorption, at different excitation wavelengths, normalized at 685 nm (Ramos and Lagorio, 2004 - Reproduced by permission of the Royal Society of Chemistry (RSC) for the European Society for Photobiology, the European Photochemistry Association, and the RSC).

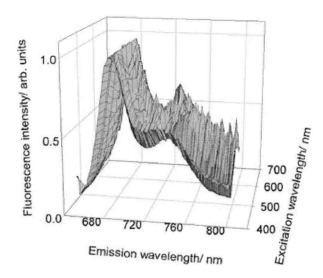

Figure 9. Emission spectra of chloroplasts isolated from *Ficus benjamina* at different excitation wavelengths, normalized at 685 nm from (Ramos and Lagorio, 2004 – Reproduced by permission of the Royal Society of Chemistry (RSC) for the European Society for Photobiology, the European Photochemistry Association, and the RSC).

Ramos and Lagorio have found that this two-flux model was reasonably good when working with groups of leaves so that no light is transmitted through them (Ramos and Lagorio, 2004).

The application of correction methods to account for re-absorption of chlorophyll fluorescence emission in leaves is subject to a number of controversies in the literature. Cordon and Lagorio analyzed and discussed comparatively three correction methods (Lagorio et al., 1998; Gitelson et al. 1998 and Agati et al., 1993), which were based on transmittance and/or reflectance measurements (Cordon and Lagorio, 2006). The method proposed by Gitelson et al (GM) gave high values for the corrected fluorescence ratio between 685 nm and 737 nm ($F_{685}/F_{737} \approx 7$ to 20 according to the different species of leaves). The two other methods-Lagorio et al. method (LM) and Agati et al. method (AM)-were found to give similar results with corrected fluorescence ratios that varied according the species between 1.4 and 2.4 about (Figure 10). Posterior studies for the abaxial faces of several dicotyledonous leaves showed that the application of the correction model LM leads to values between 2.3 to 3.5 (Cordon and Lagorio, 2007).

A model to correct light re-absorption in Granny Smith apples was also developed in literature (Ramos and Lagorio, 2006) and a value about 2 (similar to those obtained in leaves) was reported for the fluorescence ratio $F_{red}/F_{far-red}$ obtained from the corrected spectra (Figure 11).

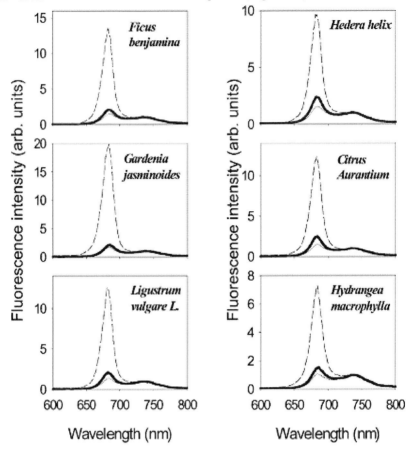

Figure 10. Fluorescence spectra for different species corrected by the detector response and by light re-absorption processes according to GM (---), AM (—), and LM (■) and normalised at 737 nm. Excitation wavelength: 468 nm. Air was used as background for the measurements of single leaves. Figure from (Cordon and Lagorio, 2007)
Reproduced by permission of the Royal Society of Chemistry (RSC) for the European Society for Photobiology, the European Photochemistry Association, and the RSC.

Further efforts to describe the true chlorophyll fluorescence spectrum emerging from chloroplasts have been made. Pedrós et al. modeled the spectral distribution of fluorescence free of re-absorption at the initial step of photosynthesis (Pédros et al., 2008). They considered the contribution of PSI and PSII to fluorescence, the stoichiometry of PSII/PSI reaction centers, the

PSII antenna size and the fluorescence lifetimes for PSII and PSI. They estimated a value of about 3.5-slightly dependent on the irradiation intensity- for the fluorescence ratio of the emission spectrum "inside a leaf".

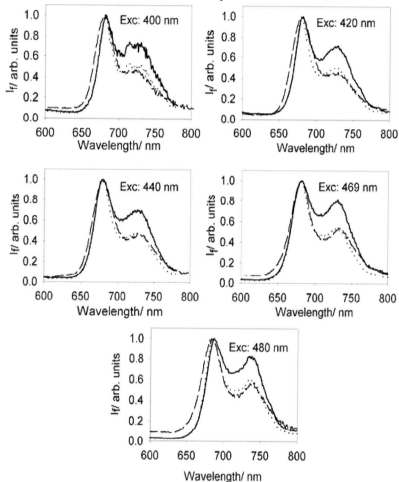

Figure 11. Experimental fluorescence spectra (corrected for the instrumental response) for whole Granny Smith apple (—), reabsorption-corrected spectra (...) and fluorescence spectra (corrected for instrumental response) for a thin layer of chloroplasts (---) at different excitation wavelengths. All spectra are normalized at 680 nm. Figure reproduced from (Ramos and Lagorio, 2006). Reproduced by permission of the Royal Society of Chemistry (RSC) for the European Society for Photobiology, the European Photochemistry Association, and the RSC.

Fluorescence spectra for dilute suspensions of chloroplasts, where no reabsorption artifacts were present, have been reported in literature. Fluorescence ratios around 3.5-3.7 have been found for these suspensions at room temperature (Franck et al., 2005; Murata et al., 1966). However an interesting point rises from the fact that chloroplasts deposited as a film on a plate display a lower fluorescence ratio (Ramos and Lagorio 2006) than chloroplasts suspended in water (Franck et al., 2005; Murata et al., 1966). This difference is remarkable especially when in both cases dilution of the samples guarantees the absence of light re-absorption processes. This experimental evidence needs some kind of debate that is not present in literature up to the moment.

It can be concluded that the true fluorescence spectra of leaves, i.e. the distribution of fluorescence emission emerging from chloroplasts in vivo (free from light re-absorption) is still a matter of controversies and further work is necessary to assess a solid base in this direction.

Chlorophyll Fluorescence in Adaxial and Abaxial Faces of Leaves

The spectral distribution of Chlorophyll fluorescence is different for the adaxial (upper) and abaxial (lower) parts of leaves. It was reported that both the experimental fluorescence intensity and the experimental fluorescence ratio $F_{red}/F_{far-red}$ were higher for abaxial than for adaxial leaves (Lang et al., 1991; Louis et al., 2006; Buschmann, 2007). These differences were attributed to the different magnitude of the re-absorption process as a result of the different chlorophyll content in both parts (Buschmann, 2007). However, Cordon and Lagorio, after applying the correction model LM to eliminate light re-absorption artifacts, reached the conclusion that the differences were not only due to differences in light re-absorption processes as there remained still a higher ratio for $F_{red}/F_{far-red}$ for abaxial leaves after correction (Cordon and Lagorio, 2007). They proposed a higher ratio PSII/PSI in abaxial leaves to account for the remaining differences after correcting for light re-absorption processes. These results are consistent with previous works reporting that chloroplasts from palisade have different photosynthetic properties than chloroplasts from spongy tissues (as those of sun and shaded plants) (Terashima and Inoue, 1985). According them, the biochemical and ultrastructural properties of the chloroplasts change gradually from sun-to shade-type with the depth from the adaxial surface. They reported a decrease in Chlorophyll a/b ratio and an increase in PSII/PSI ratio from adaxial to abaxial faces of leaves. It was reported higher Chlorophyll a/b ratio for sun type chloroplasts than for shade-type ones. Plants usually control the chlorophyll a to b ratio in order to adjust the light absorption according the intensity and the spectral distribution of irradiation light (Lichtenthaler and Burkart, 1999). Chlorophyll-a is found in every part of the photosystems but Chlorophyll-b is only found in the peripheral part of the antenna complex. Plants grown under high irradiation intensities usually display high Chlorophyll a/b ratio due to a small antenna size.

Applications of Chlorophyll Fluorescence

As we have described in the previous sections, Chlorophyll fluorescence is strongly related to the photosystem activity in natural organisms. This characteristic makes chlorophyll fluorescence analysis (either the non-variable spectral distribution or the Kautsky kinetics) a valuable tool to infer information about the physiological state of vegetation. Obviously, it opens a great deal of applications that increase everyday. Some punctual applications are briefly described below.

Forestry Health

Chlorophyll fluorescence was proved to correlate with other methods of physiological analysis of vegetation. It has become then a non-destructive indicator of disorders taking place in the photosynthetic process. It was widely used in forest's health evaluation. A review of the practical applications in forestry can be found at (Mohammed et al., 1995).

Plant Stress

The changes in the values of the photosynthetic parameters derived from Kautsky kinetics (with respect to control or healthy species) are used to study stress in plants suffering different factors. Disturbances in the photosynthetic parameters have been detected for plant exposition to high or low temperatures (Havaux, 1993; Haldimann, 1997), sun or shaded environments (Greer and

Halligan, 2001), drought (Ögren et al., 1985), air pollutants (Calatayud et al., 2002), herbicides (Dewez et al., 2002), heavy metals (Ralph and Burchett, 1998), etc. Also, variations in the fluorescence ratio during chilling (Agati et al., 2000), senescence (Cordon and Lagorio, 2007), or nutrient starvation (Yaryura et al., 2009), have been detected.

Vegetables and Fruits Monitoring during Ripening and Post-Harvesting Period

Several fruits contain chlorophylls (kiwi, banana, apple, avocado, etc.) and display typical chlorophyll-a fluorescence emission similar to that presented for leaves (Ramos and Lagorio, 2006). Even more, most of them display photosynthetic activity, during their ripening and post-harvesting periods, comparable to that originated in leaves (Khanizadeh and DeEll, 2003). This feature allows monitoring physiological disorders and the presence of stress factors from measurements of chlorophyll fluorescence. Both steady state spectra and Kautsky Kinetics are useful tools to infer information on the fruit maturity and quality evolution during the storage time (Beaudry et al., 1997; Woolf and Laing, 1996). A review about the potential of chlorophyll fluorescence in fruit breeding can be read at (Khanizadeh and DeEll, 2003).

Non-destructive methods were developed to assess Anthocyanins content in grape berries and in olive fruits by measuring Chlorophyll fluorescence excitation spectra (Agati et. al., 2005; Agati et al., 2007). Non-destructive assessment of anthocyanins, carotenoids and flavonols through Chlorophyll fluorescence measurements have been also presented in literature by Merzlyak et al. (Merzlyak et al., 2008).

Estimation of Algae Composition and Concentration

Green, blue, brown and mixed algae contain photosynthetic pigments as phycocyanobilin, phycoerythrobilin, fucoxanthin, peridinin and Chlorophyll-a. They present photosynthetic activity and fluorescence emission from Chlorophyl-a. In particular, the excitation spectrum of Chlorophyll fluorescence has been used to determine the amount of Chlorophyll and the algal group composition of phytoplankton. Recently, a multiple-fixed-wavelength spectral fluorometer was used to quantify phytoplankton biomass and community composition in estuaries (Richardson et al., 2010). Different systems based on the measurements of chlorophyll fluorescence are reported in literature for algae group detection (Ruser et al., 1999): the Algae Online Analyser was described by Beutler et al. (Beutler, 1998), the Phyto-PAM was used by Kolbowski and Schreiber (Kolbowski and Schreiber, 1995) and a Flow Cytometer was described by Hofstraat et al. (Hofstraat et al., 1991). The pigment compositions of the antenna complexes of photosystem II in algae are determined on the base of the shapes of the excitation spectra (Ruser et al., 1999).

Remote Sensing of Chlorophyll-A Fluorescence

During the last years the remote sensing of chlorophyll fluorescence has become a relevant tool to assess the physiological state of crops. Passive and active methods were developed for this purpose.

Passive methods measure fluorescence excited directly by sunlight (Smorenburg et al., 2002; Meroni and Colombo, 2006; Louis et al., 2005; Buschmann et al., 1994). The observation of solar-induced fluorescence from space is important nowadays because it can improve the knowledge of the dynamics within the biosphere. In fact, as Chlorophyll fluorescence is related to the photosynthetic activity of vegetation, the remote detection of chlorophyll fluorescence can give information on the earth´s carbon cycle that is crucial to predict climate change. (Kharuk et al.,1994). The signal due to chlorophyll fluorescence is only 1 or 2% of the absorbed light. As it is very low compared to the fraction of reflected light, the decoupling of the two signals is very complex and the passive measurements of Chlorophyll emission becomes a difficult task (Amorós-López et al. 2007). Chlorophyll fluorescence emission is extracted from reflectance measurements using the Fraunhofer Line Discrimination principle (Cendrero et al. 2009). Fraunhofer lines are dark lines in the solar spectrum due to light absorption by elements in the sun photosphere or in the earth atmosphere. In particular, A and B Fraunhofer lines (which are used in passive methods for Chlorophyll-fluorescence detection) are due to light absorption by oxygen present in the earth atmosphere (other Fraunhofer lines are due to absorption by elements present in the sun photosphere). The Fraunhofer line discrimination method (FLD) was originally described by Plascyk (Plascyk, 1975, Plascyk and Gabriel, 1975). The base for the passive method may be summarized in the following equations set. The vegetation radiance $L(\lambda)$ may be written as the sum of two contributions: the light reflected by the leaves and the fluorescence

$F(\lambda)$. The light reflected by the sample is calculated as $r(\lambda).E(\lambda)$ where $r(\lambda)$ represents the sample reflectivity and $E(\lambda)$, the solar irradiance at a given wavelength λ:

$$L(\lambda) = r(\lambda).E(\lambda) + F(\lambda) \tag{7}$$

Measuring $L(\lambda)$ and $E(\lambda)$ inside and outside the Fraunhofer lines and assuming

$$r(\lambda_{in}) \approx r(\lambda_{out}) \text{ and } F(\lambda_{in}) \approx F(\lambda_{out}) \tag{8}$$

the sample reflectivity and fluorescence can separately be obtained:

$$r = \frac{L(\lambda_{out}) - L(\lambda_{in})}{E(\lambda_{out}) - E(\lambda_{in})} \tag{9}$$

and

$$F = \frac{E(\lambda_{out})L(\lambda_{in}) - L(\lambda_{out}).E(\lambda_{in})}{E(\lambda_{out}) - E(\lambda_{in})} \tag{10}$$

However, this method assumes that F and r are constant inside and outside the Fraunhofer line, and this supposition is not true. Variations of the original FLD method have been derived. They introduce different correction factors that are comparatively discussed in (Cendrero et al., 2009) and they are named: 3FLD (Maier et al., 2003), CFLC (Gomez-Chova et al., 2006, Moya et al., 2006) and iFLD (Alonso et al., 2008).

A boreal forest in Finland was studied by remote sensing of sunlight-induced chlorophyll fluorescence using a Passive Multi-wavelength Fluorescence Detector sensor (Louis et al., 2005). With this instrument, simultaneous measurements of Chlorophyll fluorescence in the oxygen absorption bands, at 687 and 760 nm, and a reflectance index (that correlated with CO_2 assimilation), the PRI (Physiological Reflectance Index) (Evain et al., 2004; Gamon et al., 1990), were recorded.

Active methods for remote measuring of chlorophyll fluorescence, on the other hand, use a high energy LASER as irradiation light. This methodology is not restricted to daylight period as occurs for passive methods. Usually airborne blue- pulsed excimer LASERS are used. The disadvantage of this system is the narrow area sensed during each trip. This fact implies several flights to cover a big area (Lichtenthaler, 1988b). In active methods the technology LiDAR (Laser imaging detection and ranging system) is used improving the remote measurements (Cerovic et al., 1996). LiDAR is a sensing tool that can measure the distance between the sensor and the target. LiDAR was used to measure a three dimensional structure of plants and it was combined with two-dimensional Chlorophyll-a fluorescence image. This kind of research allowed correlation between physiological responses and three-dimensional plant structures (Eguchi et al., 2008).

PAM- fluorometers has also been adapted for remote sensing and non-invasive measurements of Chlorophyll fluorescence for distances up to a few meters (Ounis et al. 2001, Cerovic et al., 1996).

Conclusion

Chlorophyll fluorescence is a relevant topic to be studied in the perspective of basic and applied research related to photosynthetic organisms. Significant advances in the area have been developed since 1960 until today. Nevertheless, several physical-chemical aspects remain still under controversies. The contribution of PSI to room temperature fluorescence spectrum is still neglected by some authors and the corrections to account for light re-absorption processes seem to have not reached a full consensus in the international scientific community.

We have come a long way on this subject but still have another long stretch ahead.

Aknowledgments

The author is grateful to the University of Buenos Aires (Projects UBACyT X113, X114 and 20020100100814) and to the Agencia Nacional de Promoción Científica y Tecnológica (BID 1201/OC-AR PICT 11685 and PICT 938) for financial support.

The author wishes to thank Jonathan Lindsey for his kind assistance with the chlorophyll absorption and emission spectra in solution.

References

Agati, G.; Fusi, F; Mazzinghi, P. and Lipucci di Paola, M. A simple approach to the evaluation of the re-absorption of chlorophyll fluorescence spectra in intact leaves. *J. Photochem. Photobiol. B*. 1993, 17, 163–171.

Agati, G. Response of the in vivo Chlorophyll fluorescence spectrum to environmental factors and laser excitation wavelength. *Pure Appl. Opt.* 1998, 7, 797-807.

Agati, G.; Cerovic, Z. G. and Moya, I. The Effect of Decreasing Temperature up to Chilling Values on the in vivo F685/F735 Chlorophyll Fluorescence Ratio in Phaseolus vulgaris and Pisum sativum: The Role of the Photosystem I Contribution to the 735 nm band. *Photochem. Photobiol.* 2000, 72, 75-84.

Agati, G.; Pinelli, P.; Cortés Ebner, S; Romani, A.; Cartelat, A. and Cerovic, Z. G. Nondestructive Evaluation of Anthocyanins in Olive (Olea europaea) Fruits by in Situ Chlorophyll Fluorescence Spectroscopy. *J. Agric. Food Chem.* 2005, 53, 1354-1363.

Agati, G.; Meyer, S.; Matteini, P. and Cerovic, Z. G. Assessment of Anthocyanins in Grape (Vitis vinifera L.) Berries Using a Noninvasive Chlorophyll Fluorescence Method. *J. Agric. Food Chem.* 2007, 55, 1053-1061.

Alonso, L.; Gomez-Chova, L.; Vila-Frances, J.; Amoros-Lopez, J.; Guanter, L.; Calpe, J.and Moreno, J. Improved Fraunhofer Line Discrimination Method for Vegetation Fluorescence Quantification. Geoscience and Remote Sensing Letters, *IEEE*. 2008, 5, 620 - 624.

Amorós-López, J.; Vila-Francés, J.; Gómez-Chova, L.; Alonso, L.; Guanter, L.; del Valle-Tascón, S; Calpe, J and Moreno, J. Remote sensing of chlorophyll fluorescence for estimation of stress in vegetation. Recommendations for future missions. Geoscience and Remote Sensing Symposium. IGARSS 2007. *IEEE International*. 2007, 3769-3772.

Babani, F. and Lichtenthaler, H. K. Light-induced and age-dependent development of chloroplasts in etiolated barley leaves as visualized by determination of photosynthetic pigments, CO_2 assimilation rates and different kinds of chlorophyll fluorescence ratios. *J. Plant Physiol*. 1996, 148, 555-566.

Beaudry, R. M.; Song, J.; Deng, W.; Mir, N.; Armstrong, P. and Timm, E. Chlorophyll fluorescence a non-destructive tool for quality measurements of stored apple fruit. Proc. International Conference on Sensors for

Nondestructive Testing: *Measuring the Quality of Fresh fruits and vegetables.* 1997, 55-56.

Beutler, M. Entwicklung eines Verfahrens zur quantitativen Bestimmung von Algengruppen mit Hilfe computergestützter Auswertung spektralaufgelöster Fluoreszenzanregungsspektren. *Diplomarbeit,* Universität Kiel. 1998.

Bilger, W. and Schreiber, U. Energy-dependent quenching of dark level chlorophyll fluorescence in intact leaves. *Photosynth. Res.* 1986, 10, 303-308.

Bilger, W. and Björkman, O. Role of the xanthophyll cycle in photoprotection elucidated by measurements of light-induced absorbance changes, fluorescence and photosynthesis in leaves of Hedera canariensis. *Photosynth. Res.* 1990, 25, 173-185.

Brewster, D. On the Colour of Natural Bodies. *Trans. Roy Soc. Edinburgh.* 1834, 12, 538-545.

Briantais J. M; Vernotte C; Krause G. H. and Weis E. Chlorophyll a fluorescence of higher plants: chloroplasts and leaves. In *Light Emission by Plants and Bacteria;* Govindjee; Amesz, J and Fork D. J. (Eds.); Academic Press: New York, NY, 1986; pp 539–583.

Broglia, M. Blue-green laser-induced fluorescence from intact leaves: actinic light sensitivity and subcellular origins. *Appl. Opt.* 1993, 32, 334-338.

Buschmann, C.; Nagel, E; Szabó, K. and Kocsányi L. Spectrometer for fast measurements of in vivo reflection, absorption and fluorescence in the visible and near infrared. *Remote Sens. Environ.* 1994, 48, 18-24.

Buschmann, C. Variability and application of the chlorophyll fluorescence emission ratio red/far-red of leaves. *Photosynt. Res.* 2007, 92, 261-271.

Butler, W. L. Energy distribution in the photochemical apparatus of photosynthesis. *Ann. Rev. Plant Physiol.* 1978, 29, 345–378.

Catalayud, A; Alvarado, J. W.; Ramírez, D. J. And Barreno, E. Effects of ozone on photosynthetic CO_2 exchange chlorophyll a fluorescence and antioxidant systems in lettuce leaves. *Physiol. Plant.* 2002, 116, 308-316.

Cendrero, M. P.; Alonso, L.; Guanter, L.; Delegido, J.; Corner, A. and Moreno, J. Análisis comparativo de métodos para la medida de la fluorescencia emitida por la vegetación. In Teledetección: *Agua y desarrollo sostenible.* XIII Congreso de la Asociación Española de Teledetección. Montesinos Aranda, S. and Fernández Fornos, L. (Eds). Calatayud. Spain, 2009, pp 425-428.

Cerovic Z. G.; Goulas, Y; Gorbunov, M; Briantais, J.M.; Camenen, L and Moya, I. Fluorosensing of water stress in plants. Diurnal changes of the mean lifetime and yield of chlorophyll fluorescence measured simultaneously and at distance with a τ - LIDAR and a modified PAM-fluorimeter, in maize, sugar beet and Kalanchoë. *Remote Sens Environ.* 1996, 58, 311-321.

Cordon, G. B. and Lagorio, M. G. Re-absorption of chlorophyll fluorescence in leaves revisited. A comparison of correction models. *Photochem. Photobiol. Sci.* 2006, 5, 735–740.

Cordon, G. B. and Lagorio, M. G. Optical properties of the adaxial and abaxial faces of leaves. Chlorophyll fluorescence, absorption and scattering coefficients. *Photochem. Photobiol. Sci.* 2007, 6, 873–882.

Dewez, D.; Marchand, M.; Eullafroy, P. and Popovic, R. Evaluation of Diuron derivatives effects on Lemma gibba by using fluorescence toxicity index. *Environ. Toxicol.* 2002, 17, 493-501.

Eguchi, A.; Konishi, A.; Hosoi, F. and Omasa, K. Three-Dimensional Chlorophyll Fluorescence Imaging for Detecting Effects of Herbicide on a Whole Plant. In Photosynthesis. Energy from the Sun: 14th International Congress on Photosynthesis, Allen, J. F.;Gantt, E. ; Golbeck, J. H. and Osmond, B (Eds.); Springer: Dordrecht, *The Netherlands*, 2008, pp 577–580.

Entcheva Campbell, P. K.; Middleton, E. M.; Corp, L. A.; McMurtrey III, J. E.; Kim, M. S.; Chappelle, E. W. and Butcher, L. M. Chlorophyll fluorescence and apparent red/near-infrared reflectance of corn foliage subjected to nitrogen deficiency. In: *Proceedings of International Geoscience and Remote Sensing Symposium (IGARSS2002)*, Toronto, Canada, June 23-27. 2002.

Evain, S.; Flexas, J. and Moya, I. A new instrument for passive remote sensing: 2. Measurement of leaf and canopy reflectance changes at 531 nm and their relationship with photosynthesis and chlorophyll fluorescence. *Remote Sens. Environ.* 2004, 91, 175-185.

Franck, F.; Juneau, P. and Popovic, R. Resolution of the Photosysten I and Photosystem II contributions to chlorophyll fluorescence of intact leaves at room temperature. *Biochim. Biophys. Acta.* 2002, 1556, 239-246.

Franck, F.; Dewez, D. and Popovic, R. Changes in the Room-temperature Emission Spectrum of Chlorophyll During Fast and Slow Phases of the Kautsky Effect in Intact Leaves. *Photochem. Photobiol.* 2005, 81, 431-436.

Gamon, J. A.; Field, C. B.; Bilger, W.; Bjfrkman, O.; Fredeen, A. L. and Penuelas, J. Remote sensing of xanthophyll cycle and chlorophyll fluorescence in sunflower leaves and canopies. *Oecologia.* 1990, 85, 1-7.

Genty B.; Briantais J. M. and Baker, N. R. The relationship between the quantum yield of photosynthetic electron transport and quenching of chlorophyll fluorescence. *Biochim. Biophys. Acta.* 1989, 990, 87-92.

Gitelson, A. A.; Buschmann, C. and Lichtenthaler, H. K. Leaf Chlorophyll Fluorescence corrected for re-absorption by means of absorption and reflectance measurements. *J. Plant Physiol.* 1998, 152, 283-296.

Govindjee. Sixty-three years since Kautsky: Chlorophyll a fluorescence. *Aust. J. Plant Physiol.* 1995, 22, 131-160.

Govindjee. Chlorophyll-a Fluorescence: A Bit of Basics and History. In Chlorophyll-a Fluorescence. A Signature of Photosynthesis. Adavances in Photosynthesis and Respiration; Papageorgiou, G. C. and Govindjee (Eds); Springer: Dordrecht, *The Netherlands*, 2004, Vol. 19, pp 1-42.

Gómez-Chova, L; Alonso Chorda, L.; Amoros Lopez, J.; Vila Frances, J.; del Valle Tascon, S; Calpe, J. ; Moreno, J. Solar induced fluorescence measurements using a field spectroradiometer. Earth Observation for Vegetation Monitoring and Water Management. *AIP Conf. Proc.* 2006, 852, 274-281.

Greer, D. H. and Halligan, E. A. Photosynthetic and fluorescence light responses for kiwifruit (Actinidia deliciosa) leaves at different stages of development on vines grown at two different photon flux densities. *Austral. J. Plant Physiol.* 2001, 28, 373-382.

Hak, R.; Lichtenthaler, H. K. and Rinderle, U. Decrease of the chlorophyll fluorescence ratio F690/F730 during greening and development of leaves. *Radiat. Environ. Biophys.* 1990, 29: 329-336.

Haldimann, P. Chilling-induced changes to carotenoid composition, photosynthesis and maximum quantum yield of photosystem II photochemistry in two maize genotypes differing in tolerance to low temperature. *J. Plant Physiol.* 1997, 151, 610-619.

Hansatech FMS1 PAM Chlorophyll Fluorometer. Instruction manual. 2009.

Havaux, M. Characterization of thermal damage to the photosynthetic electron transport system in potato leaves. *Plant Sci.* 1993, 94, 19-33.

Hofstraat J. W.; de Vreeze, M. E. J.; van Zeijl, W. J. M.; Peperzak, L.; Peeters, J. C. H. and Balfoort, H. W. Flow cytometric discrimination of phytoplankton classes by fluorescence emission and excitation properties. *J. Fluoresc.* 1991, 1, 249-265.

Iriel, A.; Lagorio, M. G.; Dicelio, L. E. and San Román, E. Photophysics of Supported Dyes: Phthalocyanine on Silanized Silica. *Phys. Chem. Chem. Phys.* 2002, 4, 224-231.

Kautsky, H. and Hirsch, A. Neue Versuche zur Kohlensäureassimilation. *Naturwissenschaften.* 1931, 19, 964-964.

Kautsky, H; Appel, W and Amann, H. Chlorophyll fluorescence and carbon assimilation. Part XIII. The fluorescence and the photochemistry of plants. *Biochem. Z.* 1960, 332, 277-292.

Khanizadeh, S. and DeEll, J. R. The potential of chlorophyll fluorescence in fruit breeding. In Practical Applications of Chlorophyll Fluorescence in Plant Biology; DeEll, J. R. and Toivonen, P. M. A. (Eds); Kluwer Academic Publishers: Dordrecht, *The Netherlands*, 2003, pp 243-256.

Kharuk, V. I.; Morgun, V. N.; Theisen, A. F.; Rock, B. N. and William, D. L. Some Aspects of Chlorophyll Fluorescence Application in Remote Sensing. Geoscience and Remote Sensing Symposium. IGARSS '94. Surface and Atmospheric Remote Sensing: Technologies, Data Analysis and Interpretation. *International.* 1994, 973-975.

Kitajima, M. and Butler, W. L. Quenching of chlorophyll fluorescence and primary photochemistry in chloroplasts by dibromothymoquinone. *Biochim. Biophys. Acta.* 1975, 376, 105-115.

Kolbowski, J. and Schreiber, U. Computer controlled phytoplankton analyser based on chlorophyll fluorescence analysis using 4 different wavelengths. Xth International Photosynthesis Congress. *Photosynth. Res.* 1995, S1:10-013.

Krause, G. H. and Weis, E. Chlorophyll fluorescence and photosynthesis: the basics. *Ann. Rev. Plant Physiol. Plant Mol. Biol.* 1991, 42, 313-349.

Lakowicz, J. R. *Principles of fluorescence spectroscopy;* Springer: New York, NY, 2006; Vol. 1, pp 1-26.

Lagorio, M. G.; Dicelio, L. E.; Litter M. I. and San Román, E., Modeling of fluorescence quantum yields of supported dyes. Aluminum carboxyphthalocyanine on cellulose. *J. Chem. Soc., Faraday Trans.* 1998, 94, 419-425.

Lagorio, M. G.; San Román, E.; Zeug, A.; Zimmermann J. and Roeder, B. Photophysics on surfaces: Absorption and luminescence properties of Pheophorbide-a on cellulose. *Phys. Chem. Chem. Phys.* 2001, 3, 1524-1529.

Lang, M.; Strober, F. and Lichtenthaler, H. K. Fluorescence emission spectra of plant leaves and plant constituents. *Radiat. Environ. Biophys.* 1991, 30, 333-347.

Lehninger, A. L., Biochemistry. *The Molecular Basis of Cell Structure and Function* (2nd Edition) Worth Publishers: New York, NY, 1975, pp 599-628.

Lichtenthaler, H. K. and Buschmann, C. Photooxidative changes in pigment composition and photosynthetic activity of air-polluted spruce needles (Picea abies L.) In *Advances in Photosynthesis Research*. Sybesma, C. (Ed.); Nijhoff, M. and Junk, W. Publ:The Hague, Boston, Lancaster. 1984; vol. IV, pp 245-250.

Lichtenthaler, H. K. and Rinderle, U. The role of Chlorophyll fluorescence in the detection of stress conditions in plants. *CRC Critical Reviews in Analytical Chemistry.* 1988, 19, Suppl. I: S29-S85.

Lichtenthaler, H. K. Remote Sensing of Chlorophyll fluorescence in oceanography and terrestrial vegetation in Applications of Chlorophyll Fluorescence. In *Photosynthesis Research, Stress Physiology, Hydrobiology and Remote Sensing*; Lichtenthaler, H. K. Ed.; Kluwer Academic Publishers: Dordrecht, The Netherlands, 1988 b, pp 287-297.

Lichtenthaler, H. K.; Hák, R. and Rinderle, U. The chlorophyll fluorescence ratio F690/F730 in leaves of different chlorophyll content. *Photosynth. Res.* 1990, 25, 295-298.

Lichtenthaler, H. K.and Burkart, S. Photosynthesis and high light stress. *Bulg. J. Plant Physiol.* 1999, 25, 3-16.

Lichtenthaler, H. K; Buschmann, C. and Knapp, M. How to correctly determine the different chlorophyll fluorescence parameters and the chlorophyll fluorescence decrease ratio RFd of leaves with the PAM fluorometer. *Photosynthetica.* 2005, 43 379-393.

Louis, J.; Ounis, A.; Ducruet, J-M; Evain, S.; Laurila, T.; Thum, T.; Aurela, M.; Wingsle, G.; Alonso, L.; Pedros, R. and Moya, I. Remote sensing of sunlight-induced chlorophyll fluorescence and reflectance of Scots pine in the boreal forest during spring recovery. *Remote Sens. Environ.* 2005, 96 37–48.

Louis, J.; Cerovic, Z. G. and Moya, I. Quantitative study of fluorescence excitation and emission spectra of leaves. *J. Photochem. Photobiol. B.* 2006, 85, 65-71.

Maier, S.; Günther, K. P. and Stellmes, M. Sun-induced fluorescence: a new tool for precision farming. In *Digital Imaging and Spectral Techniques: Applications to Precision Agriculture and Crop Physiology;* VanToai, T.; Major, D.; McDonald, M.; Schepers, J. and Tarpley, L. (Eds.); American Society of Agronomy, Madison, Wisconsin, 2003, pp. 209–222.

Maxwell, K. and Johnson, G. N. J Exp Bot. 2000, 51 659-668.

Mazzinghi, P.; Agati, G. and Fusi F. *Interpretation and physiological significance of blue-green and red vegetation fluorescence International*

Geoscience and Remote Sensing Symposium (IGARSS) '94, T. I. Stein editor, (invited paper). 1994, 640-642.

Meroni, M. and Colombo, R. Leaf level detection of solar induced chlorophyll fluorescence by means of a subnanometer resolution spectroradiometer. *Remote Sens Environ.* 2006, 103, 438-448.

Merzlyak, M. N.; Bernt Melo, T. and Razi Naqvi, K. Effect of anthocyanins, carotenoids, and flavonols on chlorophyll fluorescence excitation spectra in apple fruit: signature analysis, assessment, modeling, and relevance to photoprotection. *J. Exp Bot.* 2008, 59, 349-359.

Mohammed, G. H.; Binder, W. D. and Gillies, S. L. Chlorophyll fluorescence: a review of its practical forestry applications and instrumentation. *Scand. J. For. Res.* 1995, 10, 383-410.

Moya, I; Cerovic Z, G. Remote sensing of chlorophyll fluorescence: Instrumentation and Analysis. In Chlorophyll a fluorescence: a signature of photosynthesis; Papageorgiou, G.C. and Govindjee (Eds.); Springer: Dordrecht, *The Netherlands*, 2004, pp 429-445.

Moya, I.; Ounis, A.; Moise, N. and Goulas, Y. First airborne multiwavelenght passive chlorophyll fluorescence measurements over La Mancha (Spain) fields. In *Second Recent Advances in Quantitative Remote Sensing;* Sobrino, J. A (Ed.); Publicacion de la Univ. de València, Spain, 2006, pp. 820-825.

Müller, N.J.C. Beziehungen zwischen Assimilation, Absorption and Fluoreszenz im Chlorophyll des lebenden Blattes. *Jahrbuch der wissenshaftliche Botanik.* 1874, 9, 42-49.

Murata, N; Nishimura, M and Takamiya, A. Fluorescence of chlorophyll in photosynthetic systems. II. Induction of fluorescence in isolated spinach chloroplasts. *Biochim. Biophys. Acta.* 1966, 120, 23-33.

Ögren, E and Öquist, G. Effects of drought on photosynthesis, chlorophyll fluorescence and photoinhibition susceptibility in intact willow leaves. *Planta.* 1985, 166, 389-388.

Ounis, A.; Evain, S.; Flexas, J.; Tosti, S. and Moya, I. Adaptation of a PAM-fluorometer for remote sensing of chlorophyll fluorescence. *Photosynth. Res.* 2001, 68 113-120.

Parson, W. Modern Optical Spectrosopy: With Exercises and Examples from Biophysics and Biochemistry; Springer: Dordrecht, *The Netherlands,* 2009, pp 195-197.

Pédros, R.; Moya, I.; Goulas, Y. and Jacquemoud, S. Chorophyll fluorescence emission spectrum inside a leaf. *Photochem. Photobiol. Sci.* 2008, 7, 498-502.

Peterson, R. B; Oja, V and Laisk, A. Chlorophyll fluorescence at 680 and 730 nm and leaf photosynthesis. *Photosynth. Res.* 2001, 70, 185-196.

Pfündel, E. Estimating the contribution of Photosystem I to total leaf chlorophyll fluorescence, *Photosynth. Res.* 1998, 56, 185-195.

Plascyk, J. and Gabriel, F. The Fraunhofer Line Discriminator MKII -an airborne instrument for precise and standardized ecological luminescence measurements. *IEEE Trans. Instr. Measure.* 1975, 24, 306-313.

Plascyk, J. A.The MK II Fraunhofer line discriminator (FLD-II) for airborne and orbital remote sensing of solar-stimulated luminescence. *Opt. Eng.* 1975, 14, 339-346.

Rabinowitch, E. and *Govindjee Photosynthesis*, John Wiley and Sons Inc.: New York, NY, 1969, pp 196-216.

Ralph, P. J. and Burchett, M. D. Photosynthetic response of Halophila ovalis to heavy metal stress. *Environ. Pollut.* 1998, 103, 91-101.

Ramos, M. E. and Lagorio M. G. True Fluorescence Spectra of leaves. *Photochem. Photobiol. Sci.* 2004, 3, 1063-1066.

Ramos, M. E. and Lagorio, M. G. A model considering light reabsorption processes to correct in vivo chlorophyll fluorescence spectra in apples. *Photochem. Photobiol. Sci.* 2006, 5, 508-512.

Richardson, T. L.; Lawrenz, E.; Pinckney, J. L.; Guajardo, R. C.; Walker, E. A.; Paerl, H. W. and MacIntyre, H. L. Spectral fluorometric characterization of phytoplankton community composition using the Algae Online Analyser®. *Water Res.* 2010, 44, 2461-2472.

Rodriguez, H. B.; Lagorio, M. G. and San Román, E. Rose Bengal adsorbed on microgranular cellulose. Evidence of fluorescent dimers. *Photochem. Photobiol. Sci.* 2004, 3, 674-680.

Ruser, A. Popp; P., Kolbowski, J.; Reckermann, M.; Feuerpfeil, P.; Egge, B.; Reineke, C.and Vanselow, K. H. Comparison of chlorophyll-fluorescence-based measuring systems for the detection of algal groups and the determination of chlorophyll-a concentrations, *Berichte Forsch u Technologiezentr Westküste d Univ Kiel.* 1999, Nr. 19 S. 27-38.

Smorenburg, K; Bazalgette Courrèges-Lacoste, G.; Berger M; Buschmann C; Court, A; Del Bello, U.; Langsdorf, G.; Lichtenthaler, H. K.; Sioris, C; Stoll, M-P and Visser, H. Remote sensing of solar induced fluorescence of vegetation. *Proc. SPIE* 4542. 2002 178-190.

Stokes, G. G. On the Change of Refrangibility of Light. P*hilos. T R Soc. Lond.* 1852, 142, 463-562.

Terashima, I and Inoue, Y, Vertical Gradient in Photosynthetic Properties of Spinach Chloroplasts Dependent on Intra-Leaf Light Environment. *Plant Cell Physiol.* 1985, 26 781-785.

Virgin, H. I. The Distortion of Fluorescence Spectra in Leaves by Light Scattering and Its Reduction by Infiltration. *Physiol Plant.* 1954, 7, 560-570.

Whitmarsh, J. and Govindjee. The Photosynthetic Process. In Concepts in Photobiology. Photosynthesis and Photomorphogenesis. Singhal, G. S.; Renger, G.; Sopory, S. K.; Irrgang, K. D. and Govindjee (Eds); Narosa Publishers: New Delhi and Kluwer Academic: Dordrecht, *The Netherlands*, 1999, pp.11-51.

Woolf, A. B. and Laing, W. A. Avocado fruit skin fluorescence following hot water treatments and pre-treatments. *J. Am. Soc. Hort. Sci.* 1996, 121, 147-151.

Yaryura, P.; Cordon, G.; Leon, M.; Kerber, N.; Pucheu, N.; Rubio, G.; García, A. and Lagorio, M. G. Effect of Phosphorus Deficiency on Reflectance and Chlorophyll Fluorescence of Cotyledons of Oilseed Rape (Brassica napus L.). *J. Agron. Crop. Sci.* 2009, 195, 186-196.

Zarco-Tejada, P. J; Miller J. R.; Mohammed G. H; Noland T. L. Chlorophyll fluorescence effects on vegetation apparent reflectance- I. Leaf-level measurements and model simulation. *Remote Sens. Environ.* 2000, 74, 582-595.

In: Chlorophyll
Editors: H. Le, et al.

ISBN: 978-1-61470-974-9
© 2012 Nova Science Publishers, Inc.

Chapter V

Sedimentary Chlorophyll and Pheopigments for Monitoring of Reservoir Characterized by Exclusively High Dynamism of Abiotic Conditions

L. E. Sigareva and N. A. Timofeeva
Institute for Biology of Inland Waters, Russian Academy of Sciences, pos. Borok, Nekouzskii raion, Yaroslavl oblast, Russia

Abstract

Research was made to substantiate the use of sedimentary pigments for monitoring of a trophic state of a large reservoir characterized extremely mosaic structure of bottom sediments and rare stratification. Spectrophotometric method was used to measure the concentrations of sedimentary pigments (chlorophyll a, pheopigments, total carotenoids) in the surface sediments (0–2.5 and 2.5–5.0 cm) of the Rybinsk reservoir, Russia. Bottom sediments were sampled at 6 permanent stations and from 22 to 43 stations of episodic observations at river and lake-like sites of the Rybinsk reservoir in 1993–2010. Indexes E_{480}/E_{665} and $E_{480}/(1.7E_{665acid})$ and per cent pheopigments were considered as indicators of degradation of pigment fund. Spatial (horizontal) and temporal (seasonal and long-term) distributions of sedimentary pigments in relation with water depth and temperature, Secchi depth, sediment types and concentrations of phytoplankton chlorophyll a were studied.

The concentrations of sedimentary chlorophyll a and pheopigments (Chl+Pheo) most frequently found in upper 2.5 cm sediment layer for 1993–2010 were in the range of mesortrophic and eutrophic values. The mean concentrations of Chl+Pheo for ice-free period at the separate stations varied in the range 3–284 µg/g dry matter Average of mean annual Chl+Pheo concentrations in 1993–2010 was maximum (167.4 µg/g dry matter) at ecoton site with sandy silts and clay silts, and minimum (28.4 µg/g dry matter) in lake-like part with mosaic sediments where sand was dominated.

Despite strong heterogeneity of water masses and sediment complex, the positive dependence between chlorophyll concentrations in water column and concentrations of Chl+Pheo in surface sediments of the reservoir was established. This dependence reflects the phytoplankton role in formation of productivity of bottom biotopes. Mean values of the ratio of chlorophyll content in water column to concentrations of Chl+Pheo in bottom sediments were compared with sediment accumulation rates, calculated using data of bottom sediment investigations.

The mean concentration of sedimentary Chl+Pheo in the reservoir (upper 2.5 cm layer) was calculated for two periods taking account of areas of different type sediments. For 1996–1998 it amounted to 37.0±8.5 µg/g dry matter or 15.3±2.4 mg/(m²·mm fresh matter) and for 2009–2010 – 28.1±7.5 µg/g dry matter or 10.4±3.8 mg/(m²·mm fresh matter). Decrease in sedimentary pigment content in recent years is in agreement with conception of reservoir de-eutrophication on the basis of other hydrological and hydrobiological data.

In a year with extremely hot summer and long calm weather (2010) the mean concentrations of Chl+Pheo at the stations did not differ statistically from those in normal (2009) year. However, in 2010 indexes E_{480}/E_{665} and $E_{480}/(1.7E_{665acid})$ decreased, i.e. degree of degradation of sedimentary pigment fund decreased. It was assumed that in 2010 preconditions for increase in phytoplankton productivity in the future were created.

Introduction

Among many characteristics of water ecosystems the special place belongs to primary production which, as it is known, creates a basis for functioning of all links of a trophic chain. This indicator is necessary for monitoring of productivity of the reservoirs including such stages of research as supervision, assessment and forecast. However, the experimental assessment of primary production is complicated, especially for the large reservoirs. One of indirect indicators for calculation of primary production of phytoplankton (the main source of autochthonic organic matter) is concentration of chlorophyll *a* which is quantitatively related with intensity of photosynthesis (Vinberg, 1960). One of shortcomings of estimation of primary production using chlorophyll is essential spatial and temporal variability of pigment concentration. Concentrations of plant pigments in bottom sediments are considered more conservative indicators. However, the sedimentary pigment fund of many water bodies, especially large man-made reservoirs, was not studied. Data on quantitative relationship between sedimentary pigment content and phytoplankton chlorophyll and primary production are necessary to carry out the monitoring of productivity of specific water body. Despite a number of studies in this field (Gorham et al., 1974; Flannery et al., 1982; Guillizoni et al., 1983; Adams and Prentki, 1986; Brenner and Binford, 1988; Leavitt and Findlay, 1994; Mikomägi. and Punning, 2007; Sigareva, 2006, 2010), there are not universal quantity indicators and methods. In this chapter the sedimentary pigments determined in the total extract by spectrophotometric method were analyzed. Notwithstanding its well-known shortcomings, the spectrophotometric method is still in use in limnology for analyzing pigments in the total extract because it is inexpensive and allows obtaining a large set of data on horizontal distribution of sedimentary pigments (Pincney et al., 1994). This is especially important in monitoring of water basins with a complicated relief of a bottom. The aim of research was to substantiate the use of sedimentary pigments for monitoring of a trophic state of a large reservoir characterized extremely mosaic structure of bottom sediments and rare water stratification.

Site Description

The Rybinsk Reservoir (the coordinates are 58°22'30"N 38°25'04"E) is one of the largest man-made water bodies in the Volga River cascade. It was created in 1941 and its designed level was reached in 1947. The reservoir is located in a subband of southern taiga. Its area at the normal maximum operating level is 4550 km^2, the volume is 25.4 km^3, the maximum depths is 30.4 m, the average depth is 5.6 m, and the rate of water exchange is 1.9 cycles per year. The reservoir was divided into four parts: the lake-like central main part and three river parts: Mologa, Sheksna and Volga river reaches. It is characterized by a complicated bottom relief because when the reservoir was created the river flood-lands, fluvial terraces and vast territory between Mologa and Sheksna rivers have been flooded (Rybinskoe vodokhranilishche..., 1972; Ekologicheskie problemy ..., 2001). Changes in the reservoir level depend on the runoff fluctuations in the reservoir basin and on the regime of its water resource use. Three typical periods are distinguished within the annual cycle of water level variation: spring (filling), summer–autumn (relatively steady level or insignificant drawdown), and winter (drawdown). The seasonal water level fluctuations have considerable amplitude. According to the long-term observations (1947–2004), their amplitude varied from 1.53 to 5.27 m and it averaged 3.3 m (Litvinov and Roshchupko, 2007). The averages of water transparency for vegetation periods of many years at the different reservoir parts do not exceed 1.5 m. The shallow-water zone of the reservoir represents a littoral zone up to the 4 m isobath (Gerasimov and Poddubnyi, 1999). Its area amounts to 41% of the reservoir area.

The periods of thermal and oxygen stratification of water column are rare due to water mixing by wind. The oxygen regime in general is favorable for hydrobionts. During the ice-free period, dissolved oxygen content usually is 8–9 mg/L (75–50% saturation). Under calm conditions in the middle of summer a drop of oxygen content to 1.6 mg/L (18% saturation) near the bottom at the deep-water sites can be observed. Average values of water color vary from 50 to 70° Pt–Co scale in the different parts of the reservoir (Rybinskoe vodokhranilishche..., 1972; Ekologicheskie problemy ..., 2001).

The mean thickness of sediments (14.8 cm) is maximal among all the upper Volga reservoirs (Zakonnov, 2007). The eroding activity of waves extends to a depth of 10 m. The bottom sediment distribution is mosaic. According to the investigation in 1992, the bottom sediments are presented by soils (17% of total area), sand and silty sand (55%), sandy and clayey silts (17%), peat-generated silt (2%), peaty silt (6%), peat and macrophyte deposits (3%) (Zakonnov, 1995). The mean contents of organic carbon in sand, silty sand, sandy silt, clayey silt, peat-generated silt, peaty silt correspondingly amount to 0.9, 2.5, 4.6, 7.7, 10.3, 22.4% dry matter (Zakonnov, 1993).

The ecological monitoring of the reservoir has been carried out since its creation (Ekologicheskie problemy ..., 2001). According to nutrient concentrations (nitrogen and phosphorus), the reservoir is classified as eutrophic. It was shown that the great part of the reservoir primary production is performed by phytoplankton. According to the chlorophyll *a* content in plankton, the reservoir is mesotrophic or eutrophic. In recent years the reservoir has been characterized by decrease in productivity (Devyatkin, 2003; Mineeva, 2009; Lazareva, 2010). Since 1976 increase in the average (for vegetation period) water temperature has been observed (Litvinov and Roshchupko, 2010). It was shown that long-term dynamics of phytoplankton chlorophyll *a* and primary production is in agreement with changes of solar radiation intensity (Pyrina et al., 2006). The sedimentation rate in the reservoir decreases with time (from 9.2 mm in 1941–1955 to 2.9 mm in 1992–1994). Trends of decrease in sediment accumulation rates and increase in area occupied by sands during the period of reservoir existence were revealed (Zakonnov, 1995, 2007; Zakonnov et al., 2010).

Methods

The samples of bottom sediments were collected at 6 permanent stations at the river and lake-like sites of the Rybinsk reservoir in the ice-free periods of 1993–2010. In addition, the sediment samples were collected at 43 stations of episodic observations in 1996–1998 (Sigareva et al., 1999) and 22 stations in 2008–2010 (Figure 1). The samples were taken by a stratometer. The depths at the great part of the sampling stations (4–21 m) exceeded the thickness of photosynthesis zone. The mean depths at permanent stations 1–6 correspondingly amounted to 11.5, 12.8, 6.8, 6.4, 10.6 and 12.2 m. The top layers of bottom sediments (0–2.5 and 2.5–5 cm) were analyzed. Most of the permanent sampling stations (1, 2, 5 and 6) are located within the river channels. Prevailing bottom sediment types were sandy and clayey silts at the stations 1, 2, 6 and peat-generated silt at the station 5. All the types of secondary bottom sediments of the reservoir (from sand to different silts) were found at the stations 3 and 4.

The sedimentary pigments were measured in the total acetone extract using spectrophotometric method. Double extraction of pigments from sediments with natural water content was made. The optical densities of the extracts, separated from the precipitate, before and after acidification at wave lengths corresponding to the zone of maximum light absorption by chlorophyll a (665 nm) and carotenoids (480 nm) were determined. The nonspecific absorption was measured at wavelength of 750 nm where pigments do not absorb light.

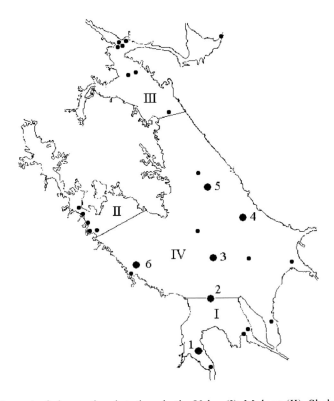

Figure 1. Layout of observational stations in the Volga (I), Mologa (II), Sheksna (III) river reaches and the lake-like main part (IV) of the Rybinsk reservoir (1 to 6 are the ordinal numbers of the permanent observational stations of the Institute for Biology of Inland Waters, RAS). Most of the permanent sampling stations are located within the river channels of Volga (st. 1, 2), Sheksna (st. 5) and Mologa (st. 6) rivers. Station 2 is ecoton that is a site in area of merge of waters of the Volga reach and main part of the reservoir.

The obtained value was subtracted from other optic parameters. The concentrations of chlorophyll *a* and pheopigments were calculated using C. J. Lorenzen equations (Lorenzen, 1967) for dry sediment, µg/g dry matter, and fresh sediment, mg/(m^2·mm), that is, per unit area (1 m^2) and 1 mm of the layer thickness. The contribution of yellow pigments was assessed from E_{480}/E_{665} – the ratio of the extract optical densities at the wavelengths of 480 and 665 nm (Burkholder et al., 1959). To eliminate the effect of pheopigments on this value, an index $E_{480}/(1.7E_{665acid})$ was used, which had the denominator other than the original in accordance with the specific coefficients of absorption of chlorophyll *a* and pheophytin *a* (Lorenzen, 1967). Indexes E_{480}/E_{665} and $E_{480}/(1.7E_{665acid})$ for sedimentary pigment fund are indicators of ratio between rates of destruction of chlorophyll and total carotenoids. The increase in indexes shows that rate of destruction of chlorophyll more than that of carotenoids.

The water transparency was determined using the Secchi disk. The water temperature was measured in the surface layer. The concentration of organic matter in bottom deposits was assessed from the loss of the mass of dry sediment on ignition at a temperature of about 600° C, and the water content was evaluated from water loss at drying at 60° C. The volume weight of the bottom sediments was calculated from the equation derived for the surface deposits in the Upper Volga reservoirs (Sigareva and Timofeeva, 2003).

Results and Discussion

The Content of Sedimentary Pigments in Relation with an Assessment of Trophic State of the Reservoir

One of approaches to an assessment of trophic state of the water bodies using sedimentary pigments was proposed on the basis of investigation of crater lakes in Germany (Möller and Sharf, 1986). The concentrations of chlorophyll+pheopigments (Chl+Pheo) in bottom sediments of these lakes do not exceed 13 in oligotrophic lakes, vary from 13 to 60 in mesotrophic and from 60 to 120 in eutrophic, and exceed 120 µg/g dry matter in hypertrophic lakes (Möller and Sharf, 1986). The gradation was used when the content of sedimentary pigments in Rybinsk reservoir was analyzed. The concentrations of sedimentary Chl+Pheo most frequently observed in upper 2.5 cm sediment layer of the reservoir for 1993–2010 were in the range of mesortrophic – hypertrophic values (Figure 2).

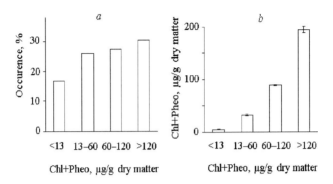

Figure 2. The distribution of the sum of chlorophyll and pheopigments (Chl+Pheo) in the samples of upper (0-2.5 cm) layer of bottom sediments of Rybinsk reservoir in 1993-2010 according to the ranges of pigment contents in oligotrophic, mesotrophic, eutrophic and hypertrophic lakes according to (Möller and Sharf, 1986). (*a*) Histogram representing the distribution of pigments and (*b*) the average pigment contents in the investigated ranges of values.

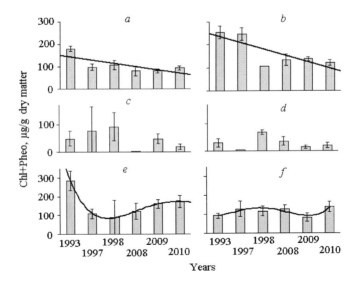

Figure 3. Long-term dynamics (1993-2010) of the sum of chlorophyll and pheopigments (Chl+Pheo) in the upper (0-2.5 cm) layer of bottom sediments of the Rybinsk reservoir. Averages of pigment content at the permanent stations: (*a–f*) correspond to stations 1–6, respectively.

The mean concentrations of Chl+Pheo for ice-free period at the separate permanent stations (1–6) varied in the range of 3–284 µg/g dry matter (Figure 3). Average Chl+Pheo concentration of many years (1993–2010) was maximum (167.4 µg/g dry matter) at the ecoton site with sandy silts and clay silts (st. 2), and minimum (28.4 µg/g dry matter) in the lake-like part with mosaic sediments where sand was dominated (st. 4).

Relationship of Sedimentary Pigment Content with Bottom Sediment Type and Water Depth

In the Rybinsk reservoirs relationship between content of sedimentary Chl+Pheo and bottom sediment type was established in 1993–1998 (Ecologiya fitoplanktona ..., 1999; Sigareva and Sharapova, 2000; Sigareva and Timofeeva, 2001, Sigareva et al., 1999). Ranked according to the increase in the concentration of pigments, given in µg/g dry matter, the bottom sediments form the series as follows: sand → silty sand → peaty silt → sandy silt → clayey silt → peat-generated silt. Such sequence was shown for all years of observations, including 2008–2010 (Table 1). Existence of close relationship between sedimentary pigments and bottom sediment type is confirmed by rather high coefficients of correlation (P<0.05) between the content of Chl+Pheo (µg/g dry matter) and the bottom sediment characteristics which reflect the sediment type: water content (r=0.74), volume mass (r= -0.69) and organic matter (r=0.56). Less close relationship between pigments and organic matter content is caused by the presence of peaty silt characterized by the maximum of organic matter and the minimum of Chl+Pheo in a unit mass of organic matter. The average content of Chl+Pheo in organic matter of different types of sediments varied from 0.11 (peaty silt) to 0.85 mg/g organic matter (sandy and clayey silts). Heterogeneity of horizontal distribution of sedimentary pigments in relation with mosaic distribution of bottom sediments should be considered when averages for the reservoir will be calculated.

Water depth is a factor which reflects complex influence of biotic and abiotic factors on the content of sedimentary pigments (Fallon and Brock, 1980; Swain, 1985; Hilton et al., 1991; Yacobi et al., 1991; Leavitt, 1993; Ostrovsky and Yacobi, 1999; Buchaca and Catalan, 2007). In the reservoirs characterized by exclusively high dynamism of abiotic conditions the relationship between water depth and content of sedimentary Chl+Pheo are strongly affected by the mosaic distribution of bottom sediments. In the Rybinsk reservoir the coefficients of correlation (P<0.05) between water depth and content of Chl+Pheo, given in mg/g organic matter (r=0.45) and µg/g dry matter (r=0.45–0.56), are rather low in comparison with lakes according to (Swain, 1985).

Table 1. The content of plant pigments in the different types of bottom sediments of the Rybinsk reservoir for the observational periods. Upper layer (0-2.5 cm)

Bottom sediment type	Chlorophyll a + pheopigments						
	µg/g of dry matter			mg/(m^2·mm fresh matter)		
	1996-1998	2009	2010	1996-1998	2009	2010
Sand	3.9±2.4	3.7±3.5	1.3±0.3	5.6±3.4	4.9±4.6	1.6±0.3
Silty sand	7.8±1.7	9.3±1.3	8.6±1.0	9.4±1.7	9.2±1.1	9.2±1.5
Peaty silt	37.8±15.3	20.5±0.0	24.3±0.0	9.7±0.3	1.8±0.0	3.7±0.0
Sandy silt	100.1±11.4	60.5±6.9	75.7±6.8	38.6±3.2	23.1±1.5	26.3±1.4
Clayey silt	151.9±19.4	105.8±12.3	123.8±18.3	42.2±4.2	27.0±2.1	28.8±2.7
Peat-generated silt	136.2±17.2	115.3±14.3	136.7±15.8	32.3±2.4	20.0±1.7	21.4±1.6

Note: Here and in Tables 3, the number of stations equaled 49 in 1996–1998 and 28 in 2009 and 2010.

The Content of Sedimentary Pigments in the Different Reaches of the Rybinsk Reservoir

Volga reach differed from other reaches higher concentrations of sedimentary pigments (Table 2). Lesser concentrations of pigments in other reaches were due to big contribution of sand and silty sand in the investigated samples. In recent years (2009–2010) the content of sedimentary pigments in the reaches has decreased in comparison with that in 1993 (Ecology fitoplanktona..., 1999) and 1997 (Table 2). The great part of the sum of Chl+Pheo was presented by products of chlorophyll degradation in all the reaches.

Indicators of Sedimentary Pigment Degradation

Indicators of sedimentary pigment degradation reflect the higher degree of pigment destruction in bottom biotopes as compared with the water mass (Leavitt, 1993; Sigareva, 2006, 2010). Fund of green pigments of deep-water sediments of Rybinsk reservoir mainly consist of products of chlorophyll degradation – pheopigments (Table 2).

Table 2. The content of the plant pigments in surface sediments (0–2.5 cm) of the reaches of the Rybinsk reservoir in different years (The averages of the mean contents at the sampling stations for vegetation periods)

Years	Chl+Pheo*, µg/g of dry matter	Pheo**, %	E_{480}/E_{665}	$E_{480}/(1.7E_{665acid})$
		Volga reach		
1997	126.1±40.6	85±3	2.28±0.13	1.47±0.08
2009	74.2±27.7	70±17	3.06±0.85	1.99±0.55
2010	98.9±27.0	85±3	2.84±0.36	1.84±0.21
		The central main part		
1997	92.3±23.8	82±2	2.32±0.16	1.51±0.11
2009	67.6±22.8	74±9	3.26±0.47	2.17±0.31
2010	94.1±24.1	82±1	2.80±0.28	1.85±0.18
		Sheksna reach		
1997	106.9±55.3	78±4	2.14±0.17	1.46±0.12
2009	50.3±13.1	72±12	2.42±0.44	1.60±0.29
2010	72.7±19.8	80±3	2.67±0.25	1.78±0.15
		Mologa reach		
1997	107.9±43.8	75±11	3.03±0.81	2.08±0.42
2009	14.7±5.2	70±17	3.48±0.97	2.27±0.63
2010	27.6±9.1	77±4	3.54±0.32	2.37±0.17

* The sum of chlorophyll *a* and pheopigments, ** relative content of pheopigments in their sum with chlorophyll *a*.

Table 3. The ratio of the yellow and green pigments in the different types of bottom sediments of the Rybinsk reservoir for the observational periods. Upper layer (0–2.5 cm)

Bottom sediment type	E_{480}/E_{665}			$E_{480}/(1.7E_{665acid})$		
	1996–1998	2009	2010	1996–1998	2009	2010
Sand	1.52±0.04	2.11±0.28	1.51±0.21	0.94±0.05	1.29±0.17	0.98±0.12
Silty sand	2.03±0.14	4.21±0.53	3.18±0.67	1.32±0.09	2.74±0.34	2.06±0.43
Peaty silt	1.68±0.41	6.70±0.0	11.10±0.0	1.17±0.33	4.85±0.0	8.40±0.0
Sandy silt	2.70±0.08	3.42±0.16	2.85±0.14	1.76±0.05	2.27±0.10	1.88±0.09
Clayey silt	2.24±0.08	2.93±0.15	2.74±0.23	1.51±0.05	1.95±0.10	1.82±0.16
Peat-generated silt	2.31±0.12	3.87±0.60	2.89±0.22	1.53±0.07	2.61±0.41	1.92±0.14

The values of ratio of the yellow to green pigments (indexes E_{480}/E_{665} and $E_{480}/(1.7E_{665acid})$) usually exceeded those for phytoplankton (1.2 and 1.1 for mesotrophic and eutrophic waters, correspondingly, according to (Mineeva, 2009)). Table 2 shows that values of the indexes corresponded to the level of sedimentary pigment content: increase in the content of Chl+Pheo was accompanied by decrease in indexes. The same tendency was revealed when the data were grouped in agreement with bottom sediment types (Table 1, 3). All the data confirm that in bottom sediments yellow pigments (total carotenoids) are preserved better, than green pigments (Hurley and Armstrong, 1991; Yacobi et al, 1991; Leavitt and Carpenter, 1990).

The Average Concentration of the Sum of Chlorophyll and Pheopigments

The average concentration of Chl+Pheo in the top 2.5–cm layer of bottom deposits of the Rybinsk reservoir, averaged with weights proportional to the areas occupied by sand and silty sand, sandy and clayey silts, peaty silt and peat-generated silt (Zakonnov, 1995), was calculated for two periods. For 1996–1998 it amounted to 37.0±8.5 µg/g dry matter or 15.3±2.4 mg/(m^2·mm fresh matter), for 2009–2010 it amounted to 28.1±7.5 µg/g dry matter or 10.4±3.8 mg/(m^2·mm fresh matter). Decrease in sedimentary pigment content in recent years is in agreement with conception of reservoir de-eutrophication on the basis of other hydrological and hydrobiological data. According to these average sedimentary pigment concentrations calculated for dry sediment (µg/g dry matter), the Rybinsk reservoir is typical mesotrophic. However, as stated above, the histogram representing the distribution of pigment concentrations shows that there were the bottom sediments of different trophic types in set of investigated samples with similar probability (Figure 2). The mean concentrations of Chl+Pheo at the sampling stations also show that the separate sites of the reservoir bottom are areas of different trophic types. Hence, the average content of Chl+Pheo calculated, taking into account horizontal distribution of sedimentary pigments and proportion of areas occupied by bottom sediments of different types, can be considered as the best indicator of the trophic state of reservoir characterized by strong heterogeneity of abiotic conditions.

Comparison of the Pigment Contents in the Bottom Sediments and Water Column

To substantiate the use of sedimentary pigments for monitoring of a trophic state of a water body, data on the ratio between indicators of productivity of water and bottom components of a water ecosystem are necessary. Data testifying to direct dependence between contents of pigments in bottom sediments and plankton can serve the proof of relationship between content of sedimentary pigments and trophic state of water body. There are some problems to assess the ratio of pigments first of all due to differences of temporal dynamics and spatial distribution of plant pigments in water and bottom biotopes.

Table 4. Contents of the sum of chlorophyll and pheopigments (Chl+Pheo) in water and in the upper (0–5 cm) layer of bottom sediments and ratio between them at the permanent stations 1–6 of the Rybinsk reservoir in 1993. Averages for free-ice period at the separate stations and for all the stations in the separate terms

Station	Period	Content of Chl+Pheo in water, mg/m²	Content of Chl+Pheo in bottom sediments, mg/(m²·mm fresh matter)	Ratio of Chl+Pheo in water to bottom sediments
	Averages at the separate stations
1	VI–X	75.5±22.3 (78)	53.8±5.2 (25)	1.3+0.3 (64)
2	VI–X	151.5±34.0 (59)	78.9±7.5 (25)	1.9±0.1 (58)
3	VI–X	98.3±23.4 (58)	15.8±3.6 (56)	10.8±4.2 (96)
4	VI–X	89.1±16.1 (48)	8.8±2.0 (60)	20.1±10.7 (141)
5	VI–X	134.9±27.8 (55)	42.6±5.8 (36)	3.7±0.9 (68)
6	VI–X	101.4±27.5 (72)	43.4±5.3 (32)	2.3±0.4 (45)
	Averages for 6 stations in the separate terms
1–6	8 VI	105.7±17.4 (40)	39.2±9.1 (57)	3.6±1.0 (67)
1–6	22 VI	124.1±20.6 (40)	46.2±16.1 (86)	19.7±13.3 (165)
1–6	6 VII	102.2±24.9 (55)	51.2±16.5 (72)	3.3±1.4 (91)
1–6	20 VII	60.8±11.8 (33)	36.3±12.0 (68)	4.9±2.9 (147)
1–6	7 IX	199.1±41.4 (51)	41.7±11.8 (69)	8.2±2.9 (87)
1–6	12 X	52.6±11.7 (50)	32.1±7.7 (53)	2.3±0.9 (86)
1–6	26 X	60.8±12.0 (48)	36.3±8.4 (57)	2.6±0.9 (88)

Note: The coefficient of variation, %, is in parentheses.

The ratio between concentrations of Chl+Pheo in the water column (mg/m^2) and upper layer (0–5 cm) sediments ($mg/(m^2 \cdot mm$ fresh matter)) was calculated using the data on seasonal dynamics of these indicators at the permanent stations (1–6). The investigated ratio varied from 0.1 to 83.0, with values of 1–3 being the most often noted. High values of the ratio are characteristic of the sites with sandy bottom (stations 3 and 4); low values were found everywhere during the seasonal recession of phytoplankton dynamics (Table 4). According to coefficients of variation, the main reasons for the variability of this parameter are the factors of the spatial distribution of pigments in the water body, but not the seasonal fluctuations. The most frequently found values of the ratio (1–3) are comparable with the average sediment accumulation rate in Rybinsk reservoir (2.9 mm/year (Zakonnov, 2007)).

The similar values of the investigated ratio of pigments were obtained by other method using the average contents of pigments in water and bottom sediments when the whole of water volume (25.4 km^3) and the total area (4550 km^2) of the reservoir were taken into account. So, the contents of Chl+Pheo amounted to 239.2 tons in water, and 64.6 tons in the upper 1-mm-thick layer of bottom sediments over the whole area, their ratio was equal to 3.7 (Sigareva, 2006). The similarity of average sediment accumulation rate with the mean ratio of Chl+Pheo concentrations in water to bottom sediments may be explained by the participation of plant suspension in the formation of sediments.

Example of Use of Sedimentary Pigments in Monitoring of the Extreme Natural Phenomena

Algal pigments in sediments are used as markers of ecosystem and climate changes (Leavitt and Findlay, 1994; Steenbergen et. al., 1994; Chen et al., 2001; Kowalewska, 2001; Buchaca and Catalan, 2007; Reuss et al., 2010). The long-term observations of pigment dynamics in the upper layer of bottom sediments allow estimating influence of abnormal environmental conditions on productivity of the Rybinsk reservoir. A role of an extreme heat was assessed using the data of 2010. The air temperature in summer of 2010 topped its characteristic values by 10–15° C and reached 35–40° C in the area of the Rybinsk reservoir.

In 2010 tendencies of spatial and seasonal changes of the water temperature of surface layer were typical of the Rybinsk Reservoir (Figure 4).

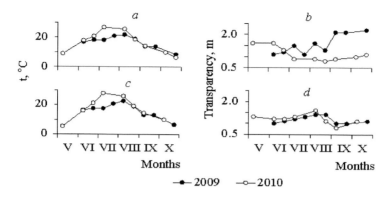

Figure 4. Seasonal dynamics of water temperature and transparency at the permanent stations of the Rybinsk reservoir in 2009 and 2010. (*a, b*) correspond to river station 1, (*c, d*) correspond to station 6 in lake-like part of the reservoir.

The greatest changes in temperature of different parts of water area were observed in May (coefficient of variation $C_V=50\%$) and in October ($C_V=11\%$), and the least changes were found in summer ($C_V=1–5\%$) when the water masses are evenly warmed. The summer water temperature rose up to 27.9 °C in 2010 and 22.8 °C in 2009.

The water transparency in 2010 (limits: 0.5–1.9 m, average value: 1.1±0.0 m) was less than in 2009 (limits: 0.7–2.7m, average value: 1.3±0.0 m). Its values were similar to most frequently observed values (Ekologicheskie problemy ..., 2001). Seasonal dynamics of a transparency at the river and lake-like stations was most differed (Figure 4). According to inverse dependence of transparency on content of suspended plankton matter, it is possible to believe that in extreme year concentration of phytoplankton chlorophyll *a* exceeded that in usual year, and it was shown most strongly in Volga reach. Water transparency of 1.1–1.3 m was observed in 1978–1995 and corresponding concentrations of chlorophyll *a* changed from 8 to 19 µg/l, i.e. in limits of mesotrophic-eutrophic values (Pyrina, et al., 2006).

Under conditions of strong warming up of water the essential increase in productivity could be expected. Positive dependence of primary production and destruction of organic matter in plankton and bottom sediments on temperature was shown earlier in research of Rybinsk reservoir (Romanenko, 1985; Devyatkin, 2003; Mineeva, 2009; Dzyuban, 2010). In 2010 the characteristics of fund of sedimentary pigments differed from those in other years. So, the maximum concentration of Chl+Pheo in the upper layer (0–2.5 cm) of bottom sediments in 2010 (297.5 µg/g of dry matter) did not reach the maximum noted for all previous period of observations (593.6 µg/g of dry matter for 1993–2009) (Sigareva and Sharapova, 2000; Sigareva and Timofeeva, 2001). In 2010 most frequently observed concentrations of sedimentary pigments are displaced in comparison with those in previous year in area of the in eutrophic and high-eutrophic values according to gradation (Möller and Scharf, 1986).

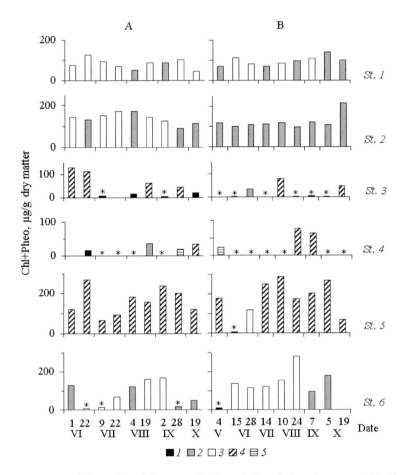

Figure 5. Seasonal dynamics of the sum of chlorophyll and pheopigments (Chl+Pheo) in the upper (0-2.5 cm) layer of bottom sediments at the permanent stations (St. 1-6) of the Rybinsk reservoir (A) in 2009 and (B) 2010. The bottom sediments types are (*1*) sand or silty sand, (*2*) sandy silt, (*3*) clayey silt, (*4*) peat-generated silt, (*5*) peaty silt. * Low or track concentrations.

Seasonal dynamics of sedimentary pigments at the stations of water bodies is usually characterized by peaks (in spring or early in summer and in autumn) following peaks of phytoplankton abundance (Rippey and Jewson, 1982; Ecologiya fitoplanktona ..., 1989; Hilton et al., 1991). In early studies of fund of sedimentary pigments of the Rybinsk reservoir it was revealed that pattern of seasonal variability of sedimentary Chl+Pheo was affected by heterogeneity of spatial distribution of types of bottom sediments (Sigareva, 2010). In the given research it was shown that heterogeneity of bottom sediments was a principal cause of the strongest seasonal changes of concentration of Chl+Pheo at the rather shallow stations (st. 3 and 4) in the lake-like central main part of the reservoir (Figure 5). Seasonal dynamics of Chl+Pheo in bottom sediments of the same type was characterized by peaks in typical periods – in early summer and in autumn. Peculiarities of seasonal changes in 2010 were a shift of autumn peak to later terms and an excess of autumn maximum over spring peak. Besides, in 2010 the maxima of concentrations of sedimentary Chl+Pheo at the majority of stations were more than those in previous year. The highest coefficients of variation of Chl+Pheo (μg/g of dry matter) at separate stations were in extreme year. So, at the stations 3 and 4 the coefficients of variation reached 99 and 113% in 2009 and 144 and 161% in 2010.

Values of coefficient of variation characterizing heterogeneity of spatial distribution of sedimentary pigments were less (up to 106%) than that characterizing seasonal change at the separate stations (up to 161%). Average values of coefficient of variation (spatial distribution) amounted to 75% in 2009 and 86% in 2010. Hence, spatial variability of Chl+Pheo in extreme year was more than that in usual year. The possible reason of it was decrease of hydrodynamic activity in calm conditions that leaded to increase of heterogeneity of phytoplankton distribution and sedimentation.

Average pigment concentrations at six permanent sampling stations for the vegetation period of 2010 did not differ significantly from those for 2009, despite distinctions of seasonal dynamics of Chl+Pheo in these years (Figure 3, 5). Average concentrations of sedimentary Chl+Pheo in 2010 were essentially less than those in 1993 (Sigareva, 2010) (Figure 3). The concentrations in 2010 increased in comparison with averages of annual concentrations for 1993–2009 only at stations (5 and 6) influenced by water masses of the Sheksna and Mologa rivers and they decreased at the other stations. The great variability of average concentrations in long-term and seasonal aspects was characteristics of stations with mosaic type of bottom sediments. In 2010 the pigment content in silts was somewhat more than that in previous year, but it did not reach values obtained for silts for 1996–1998 (Table 1) (Sigareva and Sharapova, 2000).

The average concentration of Chl+Pheo in the top 2.5–cm layer of bottom deposits of the Rybinsk reservoir, averaged with weights proportional to the areas occupied by sand and silty sand, sandy and clayey silts, peaty silt and peat-generated silt (Zakonnov, 1995), amounted to 26.6±7.6 µg/g of dry matter or 10.8±2.7 mg/(m^2·mm fresh matter) in 2009. In 2010 it amounted to 29.8±8.3 µg/g dry matter or 10.4±3.8 mg/(m^2·mm fresh matter). These results show that the content of sedimentary pigments did not increase significantly in 2010 and the trophic state of the water body did not change, i.e. it remained mesotrophic according to the gradation (Möller and Sharf, 1986).

The relative content of pheopigments in year with abnormal heat summer (2010) was similar to that in previous years (Table 2). However, in 2010 the proportion of the yellow and green pigments decreased: indexes E_{480}/E_{665} and $E_{480}/(1.7E_{665acid})$ in 2010 were less than those in 2009 (Table 2, 3). Their values somewhat approached those for phytoplankton. According to research (Furlong and Carpenter, 1988; Sun et al., 1993; Villanueva and Hastings, 2000; Datsenko, 2007; Martynova, 2010), it can be believed that in 2010 there were more favorable conditions for preservation of organic matter and pigments in surface sediments due to decrease of oxygen concentration in near-bottom layers of water owing to long calm weather and stratification.

One of indicators of dynamics of reservoir productivity is the ratio between concentrations of Chl+Pheo in surface layers of bottom sediments. In the Rybinsk reservoir there is direct correlation between the content of Chl+Pheo (µg/g of dry matter) in layers of bottom sediments of 0–2.5 and 2.5–5 cm (in 2009 r=0.73, in 2010 r=0.85). In these layers microbial processes are most active (Romanenko, 1985). The relation reflects in most cases decreases of destruction processes and content of vegetative pigments in the bottom layers. In 2010 the ratio between concentrations of Chl+Pheo in compared layers was more than that in 2009. It testifies that in 2010 accumulation of pigment in the upper (0–2.5 cm) layer increased and degree of their degradation decreased (Table 5).

Conditions of abnormal year did not change character of the relations between the distribution of sedimentary Chl+Pheo and abiotic factors. However, the coefficient of correlation of pigment content with water depth increased and amounted to 0.56 in 2010 whereas in 2009 and 1996–1998 it was equal to 0.45 (Sigareva and Sharapova, 2000). Probably, it was caused by decrease of influence of hydrodynamic activity of water masses on the plankton distribution in calm conditions. A close positive correlation between the Chl+Pheo concentrations and water content remained. The correlation coefficient reached 0.74 in 2009 and 0.78 in 2010.

Table 5. Mean ratios between the content of Chl+Pheo (μg/g of dry matter) in layers of bottom sediments of 0–2.5 and 2.5–5 cm at the permanent stations of the Rybinsk reservoir for ice-free periods in 2009 and 2010

Year	Ratio at the stations (1–6)						
	1	2	3	4	5	6	1–6
2009	1.20± 0.07	1.28± 0.09	2.44± 0.92	0.63± 0.24	1.66± 0.44	1.17± 0.22	1.45±0.17 (n=45)
2010	1.78± 0.33	1.34± 0.19	1.88± 0.34	1.28± 0.26	1.75± 0.44	1.94± 0.59	1.67±0.15 (n=49)

Note: The main factor of changes of the ratio at the station 3 and 4 is mosaic distribution of bottom sediments.

Thus, comparison of content of sedimentary pigment in abnormal and usual years allowed estimating influence of extreme heat on production properties of bottom sediments of the reservoir. Despite similar dynamics of water temperature at the sampling stations, different changes of production indicators were revealed. Increase in the water temperature did not significantly affect the average content of sedimentary Chl+Pheo in 2010. In 2009–2010 decrease in concentrations of sedimentary pigments in comparison with those in 1993–1998 was found. This is in agreement with idea of de-eutrophication of the reservoir. However, the trophic state has not changed and the reservoir remains a mesotrophic water body. Besides, increase in water temperature led to changes of dynamics and ratio between concentrations of green and yellow plant pigments in bottom sediments. There are bases to believe that in 2010 there were more favorable conditions for preservation of sedimentary pigments and, naturally, organic matter. In the future it is possible to expect an increase in destruction of organic matter of bottom sediments and release of nutrients (mainly N and P) from sediments to water column that will lead to increase in phytoplankton productivity because there is the direct positive relation between concentrations of sedimentary Chl+Pheo and biogenic elements (Timofeeva and Sigareva, 2004). Hence, after the long period of de-eutrophication of the Rybinsk reservoir noted on various hydrobiological and hydrological indicators (Devjatkin, 2003; Zakonnov, 2007; Mineeva, 2009; Lazareva, 2010), the activization of eutrophication of its ecosystem is possible.

Conclusion

The results have shown the unity of biotic and abiotic factors influencing the productivity of the reservoir ecosystem. Spatial (horizontal) and temporal dynamics of sedimentary plant pigments may be used for monitoring of trophic characteristics of reservoirs with unstable hydrological regime and mosaic distribution of bottom sediments. Sedimentary pigments in reservoirs are biomarkers of environmental changes including global climate warming. The response of fund of sedimentary pigments to abnormal high temperature depended on peculiarity of biotopes and dynamics of abiotic conditions. The response consisted in decrease in the ratio of yellow to green pigments, increase in ratio of concentrations of Chl+Pheo in the upper to more low located layers, although the trophic state of the reservoir has not changed. The average concentrations of Chl+Pheo in surface sediments of the Rybinsk reservoir throughout the investigation period (1993–2010) were in limits of mesotrophic values.

References

Adams, M. S. and Prentki, R. T. (1986). Sedimentary pigments as an index of the trophic status of Lake Mead. *Hydrobiologia*, vol. 143, pp. 71–77.

Brenner, M. and Binford, M. W. (1988). Relationships between concentrations of sedimentary variables and trophic state in Florida lakes. *Can. J. Fish. Aquat. Sci.,* vol. 45, pp. 294–300.

Buchaca, T. and Catalan, J. (2007). Factors influencing the variability of pigments in the surface sediments of mountain lakes. *Freshwater Biology*, vol. 52, no. 7, pp. 1365–1379.

Burkholder, P. R., Burkholder, L. M., and Rivero J. A. (1959). Chlorophyll "a" in some corals and marine plants. *Nature,* vol. 183, no. 4671, pp. 1338–1339.

Chen, N., Bianchi, T. S., McKee, B. A., and Bland J. M. (2001). Historical trends of hypoxia on the Louisiana shelf: application of pigments as biomarkers. *Organic Geochemistry*, vol. 32, no.4, pp. 543–561.

Datsenko, Yu. S. (2007). *Evtrofirovaniye vodokhranilishch. Gidrologogidrokhimicheskiue aspecty* (Water reservoir eutrophication). Moscow: GEOS. (In Russian).

Devyatkin, V. G. (2003). *Structure and productivity of littoral algocenoses in Upper Volga reservoirs,* Extended abstract of doctoral (Biol.) dissertation. Moscow: Mosk. Gos. Univ., 43 p. (In Russian).

Dzyuban, A. N. (2010). *Destruktsiya organicheskogo veshchestva i tsikl metana v donnykh otlozheniyakh vnutrennikh vodoyomov* (Destruction of organic matter and the methane cycle in bottom sediments of inland waterbodies). Yaroslavl: Print house. (In Russian).

Ekologicheskie problemy Verkhnei Volgi (Environmental problems of the Upper Volga River) (2001). Yaroslavl: Yaroslav. Gos. Tekhn. Univ. (In Russian).

Ekologiya fitoplanktona Kujbyshevskogo vodokhranilishcha (Ecology of phytoplankton of Kujbyshev reservoir) (1989). Leningrad: *Nauka.* (In Russian).

Ekologiya fitoplanktona Rybinskogo vodokhranilishcha (Ecology of phytoplankton from Rybinsk reservoir) (1999). *Togliatti.* (In Russian).

Fallon, R.D. and Brock, T. D. (1980). Planktonic blue-green algae: production, sedimentation and decomposition in Lake Mendota, Winconsin. *Limnol. Oceanogr.,* vol. 25, no. 1, pp. 72–88.

Flannery, M. S., Snodgrass, R. D., and Whitmore, T. J. (1982). Deepwater sediments and trophic conditions in Florida lakes. *Hydrobiologia,* vol. 92, pp. 597–602.

Furlong, E. T. and Carpenter, R. (1988). Pigment preservation and remineralization in oxic coastal marine sediments. *Geochim. Cosmochim. Acta,* vol. 52, pp. 87–99.

Gerasimov, Yu. V. and Poddubnyi, S. A. (1999). *Rol' gidrologicheskogo rezhima v formirovanii skoplenii ryb na melkovod'yakh ravninnykh vodokhranilishch* (The role of hydrological regime in the formation of fish accumulation in the shallows of lowland Reservoirs). Yaroslavl: Izd. YaGTU. (In Russian).

Gorham, E., Lund, J. W. G., Sanger, J. E., and Dean, W. E. (1974). Some relationships between algal standing crop, water chemistry, and sediment chemistry in the English lakes. *Limnol. Oceanogr.,* vol. 19, no. 4, pp. 601–617.

Guilizzoni, P., Bonomi, G., Galanti, G. and Ruggiu, D. (1983). Relationship between sedimentary pigments and primary production; evidence from core analyses of twelve Italian lakes. *Hydrobiologia,* vol. 103, no. 1, pp. 103–106.

Hilton, J., Lishman, J. P., Carrick, T. R., and Allen P. V. (1991) An assessment of the sources of error in estimations of bulk sedimentary pigment concentrations and its implications for trophic status assessment. *Hydrobiologia*, vol. 218, no. 3, pp. 247–254.

Hurley, J. P. and Armstrong, D. E. (1991). Pigment preservation in lake sediments: a comparison of sedimentary environments in Trout Lake, Wisconsin. *Can. J. Fish. Aquat. Sci.,* vol. 48, no. 3, pp. 472-486.

Kowalewska, G. (2001). Algal pigments in Baltic sediments as markers of ecosystem and climate changes. *Climate Research*, vol.18, no.1–2, pp. 89–96.

Lazareva, V. I. (2010). *Struktura I dinamika zooplanktona Rybinskogo vodokhranilishcha* (Zooplankton structure and dynamics in the Rybinsk reservoir). Moscow: KMK Scientific Press Ltd. (In Russian).

Leavitt, P. R. (1993). A review of factors that regulate carotenoid and chlorophyll deposition and fossil pigment abundance. *J. Paleolimnol.*, vol. 9, pp. 109–127.

Leavitt, P. R. and Carpenter, S. R. (1990). Aphotic pigment degradation in the hypolimnion: Implications for sedimentation studies and paleolimnology. *Limnol. Oceanogr.*, vol. 35, no. 2, pp. 520–534.

Leavitt, P. R. and Findlay, D. L. (1994). Comparison of fossil pigments with 20 years of phytoplankton data from eutrophic Lake 227, Experimental Lakes Area, Ontario. *Can. J. Fish. Aquat. Sci.*, vol. 51, pp. 2286–2299.

Litvinov, A. S. and Roshchupko, V. F. (2007). Long-term and seasonal water level fluctuations of the Rybinsk reservoir and their role in the functioning of its ecosystem. *Water Resources*, vol. 34, no. 1, pp. 27–34.

Litvinov, A. S. and Roshchupko, V. F. (2010). Long-term variations of elements of hydrometeorological regime of the Rybinsk reservoir. *Russian Meteorology and Hydrology,* vol.35, no.7, pp. 483–489.

Lorenzen, C. J. (1967). Determination of chlorophyll and phaeopigments: spectrophotometric equations. *Limnol. Oceanogr.*, vol. 12, no. 2, pp. 343–346.

Martynova, M. V. (2010). *Donnye otlozheniya kak sostavlyayushchaya limnicheskikh ekosistem* (Bottom sediments as a component of limnical ecosystems). M: *Nauka*. (In Russian).

Mikomägi, A. and Punning J.-M. (2007). Fossil pigments in surface sediments of some Estonian lakes. *Proc. Estonian Acad. Sci. Biol. Ecol.*, vol. 56, no. 3, 239–250.

Mineeva, N. M. (2009). *Pervichnaya productsiya planktona v vodokhranilishchakh Volgi* (Plankton primary production in the Volga River reservoirs). Yaroslavl: Print House. (In Russian).

Möller, W. A. A. and Scharf, B. W. (1986). The content of chlorophyll in the sediment of the volcanic maar lakes in the Eifel region (Germany) as an indicator for eutrophication. *Hydrobiologia,* vol. 143, pp. 327–329.

Ostrovsky, I. and Yacobi, Y. Z. (1999). Organic matter and pigments in surface sediments: possible mechanisms of their horizontal distributions in a stratified lake. *Can. J. Fish. Aquat. Sci.*, vol. 56, pp.1001–1010.

Pincney, J., Papa, R., and Zingmark, R. (1994). Comparison of high-performance liquid chromatographic, spectrophotometric, and fluorometric methods for determing chlorophyll *a* concentrations in estuarine sediments. *J. Microbiol. Methods*, vol. 19, pp. 59–66.

Pyrina, I. L., Litvinov, A. S., Kuchai, L. A., Roshchupko, V. F. and Sokolova, Ye. N. (2006). Long-term changes in phytoplankton primary production in the Rybinsk reservoir caused by effect of climatic factors. In: A. F. Alimov, and V. V. Bulyon (Eds.), *Sostoyanie I problemy produktsionnoi gidrobiologii* (State and problems of production hydrobiology) (pp. 36–46). Moscow: KMK. (In Russian).

Reuss, N., Leavitt, P. R., Hall, R. I., Bigler, C., Hammarlund, D. (2010). Development and application of sedimentary pigments for assessing effects of climatic and environmental changes on subarctic lakes in northern Sweden. *J. Paleolimnol.*, vol. 43, no. 1, pp. 149–169.

Rippey, B. and Jewson, D. H. (1982). The rates of sediment-water exchange of oxygen and sediment bioturbation in Lough Neagh, Northern Ireland. *Hydrobiologia,* vol. 92, pp. 377–382.

Romanenko, V. I. (1985). Mikrobiologicheskie protsessy produktsii i destruktsii organicheskogo veshchestva vo vnutrennikh vodoyomakh (Microbiological processes of production and destruction of organic matter in inland water bodies). Leningrad: *Nauka*. (In Russian).

Rybinskoe vodokhranilishche i ego zhisn' (Rybinsk reservoir and its life) (1972). Leningrad: Nauka. (In Russian).

Sigareva, L. E. (2006). Formation and transformation of a plant pigment pool in water bodies of the Upper Volga River basin, Extended Abstract of Doctoral (Biol.) *Dissertation*, Moscow: Mosk. Gos. Univ., 47 p. (In Russian).

Sigareva, L. E. (2010). The chlorophyll content in water and bottom sediments of the Rybinsk reservoir. *Inland Water Biology*, vol. 3, no. 3, pp. 240–248.

Sigareva, L. E. and Sharapova, N. A. (2000). Estimation of bulk sedimentary pigment concentrations in Rybinsk reservoir, Upper Volga. Russia. In: *Osyora holodnykh regionov. Ch. III. Gidrogeohimicheskie voprosy:*

Mater. konf. (Lakes of cold regions. P. III. Hydrogeochemical questions. Proc. Conf.) (pp. 5–15). Yakutsk: Yakutsk. Gos. Univ.

Sigareva, L. E, Sharapova, N. A., and Bashkin, V. N. (1999). Phytopygment indexes for assessment of water body state and loading at water ecosystems. Sedimentary pigments in reservoirs. In: The calculation and mapping of critical loads for air pollutants relevant to the UN/ECE convention on long-range transboundary air pollution. *Proceedings of the Second Training Workshop*, IBBP RAS, Pushchino, 1999. (pp. 91–100). Moscow: POLTEX.

Sigareva, L. E. and Timofeeva, N. A. (2001). Phyto-pigments in bottom sediments as indicators of the trophic state of Upper Volga reservoirs. *Probl. Region. Ekol.*, no. 2, pp. 23–35. (In Russian).

Sigareva, L. E. and Timofeeva, N. A. (2003). Plant pigments in the Ivankovo reservoir silts as indicators of destruction processes. *Water Resources*, vol. 30, no. 3, pp. 315–324.

Steenbergen, C. L. M., Korthals, H. J., and Dobrynin, E. G. (1994). Algal and bacterial pigments in non-laminated lacustrine sediment: Studies of their sedimentation, degradation and stratigraphy. *FEMS Microbiol. Ecol.*, vol. 13, no.4, pp. 335–351.

Sun, M.-Y., Lee, C., and Aller, R. C. (1993). Anoxic and oxic degradation of C^{14}–labeled chloropigments and a C^{14}-labeled diatom in Long Island Sound sediments. *Limnol. Oceanogr.*, vol.38, no. 7, pp. 1438–1451.

Swain, E. B. (1985). Measurement and interpretation of sedimentary pigments. *Freshwater Biol.*, vol.15, pp. 53–75.

Timofeeva, N. A. and Sigareva, L. E. (2004). Relationships between the concentrations of phyto-pigments, nitrogen, and phosphorus in bottom deposits of water reservoirs. *Water Resources*, vol. 31, no. 3, pp. 303–306.

Villanueva, J. and Hastings, D. W. (2000). A century-scale record of the preservation of chlorophyll and its transformation products in anoxic sediments. *Geochimica et Cosmochimica Acta*, vol. 64, no. 13, pp. 2281–2294.

Vinberg, G. G. (1960) Pervichnaya produktsiya vodoyomov (Primary Production of water bodies). Minsk: Izd-vo Akad. *Nauk, BSSR*. (In Russian).

Yacobi, Y, Z., Mantoura, R. F. C, and Lewellyn, C. A.. (1991). The distribution of chlorophylls, carotenoids and their breakdown products in Lake Kineret (Israel) sediments. *Freshwat. Biol.*, vol. 26, no.1, pp. 1–10.

Zakonnov, V. V. (1993). Accumulation of biogenic elements in bottom sediments in the Volga reservoirs. In: *Organicheskoe beshchestvo donnykh otlozhenij volzhskikh vodokhranilishch* (Organic matter of bottom sediments in the Volga reservoirs) (pp.3–16). Saint Petersburg: Gidrometeoisdat. (In Russian).

Zakonnov, V. V. (1995). Space and time heterogeneity in the distribution and accumulation of bottom sediments in the Upper Volga reservoirs. *Vodn. Resur.*, vol. 22, no. 3, pp. 362–371. (In Russian).

Zakonnov, V. V. (2007). Sedimentation in Volga chain reservoirs, Extended Abstract of Doctoral (Geogr.) *Dissertation,* Moscow: IG RAN, 42 p. (In Russian).

Zakonnov, V. V., Poddubnyi, S. A., Zakonnova, A. V., and Kas'yanova, V. V. (2010). Sedimentation in variable backwater zones of Volga chain reservoirs. *Water Resources*, vol. 37, no. 4, pp. 462–470.

In: Chlorophyll
Editors: H. Le, et al.

ISBN: 978-1-61470-974-9
© 2012 Nova Science Publishers, Inc.

Chapter VI

Medicinal Uses of Chlorophyll: A Critical Overview

V. K. Mishra,[1] R. K. Bacheti[2] and Azamal Husen[3]*

[1]Department of Biotechnology, Doon (P.G.) Paramedical College,
Dehra Dun-248001, India
[2]Department of Chemistry, Graphic Era University,
Dehra Dun-248001, India
[3]Department of Biology, Faculty of Natural and Computational Sciences, University of Gondar, Gondar, Ethiopia

Abstract

Reports on traditional medicinal uses of chlorophyll in alternative forms of medicine are known since ages. Now-a-days chlorophyll has been used in the field of medicine as remedy and diagnostics. Chlorophyll molecules are used in pharmacy as photosensitizer for cancer therapy. Their roles as modifier of genotoxic effects are becoming increasingly important, besides these it being known to have multiple medicinal uses. Chlorophyll has its place in modern medicine. Here, we present a review of recent developments in medicinal uses of chlorophyll. This article enumerates therapeutic claims of chlorophyll as drugs based on

---

*Department of Biology, Faculty of Natural and Computational Sciences, University of Gondar, P.O. Box 196, Gondar, Ethiopia. Email: adroot92@yahoo.co.in.

investigative findings of modern science. A brief overview of research and developments of medicinal uses of chlorophyll will be presented in this review along with challenges of potential applications of chlorophyll and its derivatives as chemotherapeutic agents.

Keywords: Chlorophyll, medicine, genotoxity, photosensitizer

Abbreviations

CHL	Chlorophyllin
ROS	Reactive oxygen species
PDT	Photodynamic therapy
PSMA	Prostrate-specific membrane antigen
ALA	Aminolevulinic acid
CDK	Cyclin dependent kinase

1. Introduction

Natural products have been the most important source of drugs. Throughout history, these products have been used as important source of anticancer and chemopreventive agents. Many natural products from our daily consumption of fruits, vegetables, tea beverages whose active ingredients have potential health benefits. Recently, their uses are becoming increasingly popular as evident from the sales of food supplements/functional foods which is growing at an amazing proportion, $4.59 billion for 2006 and $4.79 billion for 2007 (Knasmüller et al., 2008). Despite, growing body of epidemolgical and investigative findings supporting health claims of dietary supplements, there is urgent need to ensure consumer concern about their efficacy and potentiality as medicine . Among several dietary phytochemicals, chlorophyll being most ubiquitous natural pigments with physiological effects to cure of chronic diseases, such as some forms of cancer.

The chlorophyll and its derivatives have long history in traditional medicine ((Esten and Dannin, 1950; Kephart, 1955), and also various therapeutic uses including wound healing (Dashwood, 1997), anti-inflammatory agent (Bower, 1947; Larato et al., 1970), internal deodorant (Young et al., 1980). Although these applications illustrate various medicinal uses of chlorophyll but interestingly recent research works are more focused on its role as potent anti-mutagen and anti-carcinogen (Dashwood, 1997, 2002, Egner et al., 2001, 2003), and also as photosensitizer in photodynamic therapy (Henderson et al., 1997; Park, 1989; Li, et al., 2005). The intent of present article is aimed at providing better understanding of science based health claims of chlorophyll.

2. Chemotherapeutic Potential of Chlorophyll

2.1. Chlorophyll and Its Derivatives

Chlorophyll has a porphyrin ring similar to that of heme in hemoglobin, although the central atom in chlorophyll is magnesium instead of iron (Figure 1). Chlorophyllin is a semi-synthetic mixture of sodium copper salts derived from chlorophyll. During the synthesis of chlorophyllin, the magnesium atom at the center of the ring is replaced with copper and the phytol tail is lost. Unlike natural chlorophyll, chlorophyllin is water-soluble. Although the content of different chlorophyllin mixtures may vary, two compounds commonly found in commercial chlorophyllin mixtures are trisodium copper chlorin e_6 and disodium copper chlorin e_4 (Figure 2).

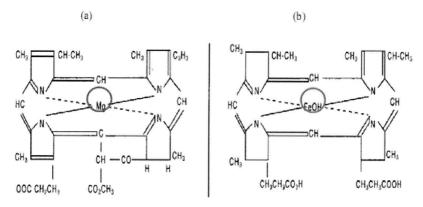

Figure 1. Molecular structure of (a) chlorophyll and (b) red blood cell.

Table 1. Chlorophyll and its derivatives used in medicine

Natural chlorophyll	Chlorophyll a, b, c, d, e
Metal free chlorophyll derivatives	Pheophytin, Pyropheophytin
Metallochlorophyll derivatives	Zn-Pheophytin
	Zn-pyropheophytin
	Chlorophyllide
	Pheophorbide
	Cu(II)chlorin e 4
	Cu-chlorin e 6
	Cu-chlorin e 4 ethyl ester

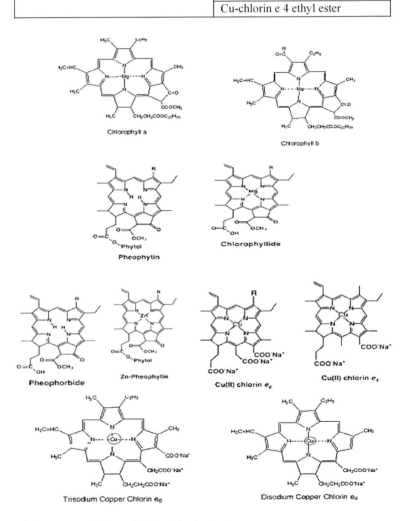

Figure 2. Structure of chlorophyll and its derivatives.

An excellent account of structure of chlorophyll and its derivatives, stability, bioavalability and their cancer preventing activity has been reviewed by Ferruzzi and Blakesle (2007). Chlorophyllin as been extensively studied for its effect in animal/human, and also utilized as food grade colorant in Europe, Asia and to a more limited and growing extent in United States (Ferruzzi and Blakesle, 2007). Some of the important chlorophyll and its derivatives are listed in Table 1.

2.2. Potential Mechanism of Action of Chlorophyll

Chlorophyll derivatives after release from the plant food matrix, natural chlorophyll (CHL) derivatives are exposed to the acidity of the gastric digestion resulting in conversion to respective metal-free pheophytins (PHE). Following digestive degration of commercial chlorophyll derivatives, they are absorbed by intestinal cells and finally passes into blood circulation (Egner, 2000, Ferruzzi et al., 2002). Chlorophyll and its derivatives act through variety of mechanisms which include: (i) antioxidant activity; (ii) modifier of genotoxic effect; (iii) inhibition of cytochrome P450 enzymes; (iv) induction of phase II enzymes; (v) increased level of gluta-thione S-transferase; (v) cell differentiation, cell arrest and apoptosis.

2.2.1. Antioxidant Effect

The major source of reactive oxygen species (ROS) is electron leakage from the mitochondrial electron transport chain, which then reacts with molecular oxygen forming ROS. ROS includes free radical such as superoxide ($O_2 \cdot -$) and hydroxyl radical ($OH \cdot$) and non-radical species such as hydrogen peroxide (H_2O_2). These free radicals set chain reaction of free radical formation when they interact with another molecule. High concentration of ROS causes oxidative damage to bio-molecules such as lipids, proteins and nucleic acids, leakage of electrolytes via lipid peroxidation, which results in the disruption of the cellular metabolism. Antioxidants act as an electron sink that neutralizes free radicals either through preventing free radical formation (preventive antioxidants) or preventing free radical chain propagation. Free radicals have been implicated to play an important role in development several diseases (Yoshikawa et al., 2000; Devasagayam et al., 2004; Knasmüller et al. 2008), which include some forms of cancer, neurological disorders, inflammatory diseases, dermatitis, tissue damage and sepsis, cardiovascular ailments (Elahi and Matata, 2006; Lefer and Granger 2000), and rheumatoid arthritis, idiopathic infertility (Agarwal et al., 2006; Pasqualotto et al., 2001), decreased immune function, several diseases of ageing (Von et al., 2004). There are contradictory views about ROS and cancer-one suggesting increased level of ROS causes cancer formation and proliferation while other opined that ROS may kill cancer cells (Schumacker, 2006).

Dietary chlorophyll derivative has ability to scavange long lived free radicals, such as 1,1-diphenyl-2-picrylhydrazyl (DPPH) and 2,2'-azino-bis-(3-ethylbenzothiazoline-6-sulfonate) (ABTS) (Ferruzi et al., 2002; Lanfer-Marquez et al., 2005). Natural chlorophyll a and b exhibited lower antioxidant activity than metal-free derivative (chlorins, pheophytins, and pyropheophytins), however metallo-derivatives (Mg-chlorophylls, Zn-pheophytins, Zn-pyropheophytins, Cu-pheophytina, and Cu-chlorophyllins) have highest antioxidant activity (Lanfer-Marquez et al., 2005). Chlorophyll and derivatives have potent antioxidant and radioprotective effects *in vitro* and *in vivo*. They inhibit lipid peroxidation (Sato et al., 1983, 1984, 1985), protein oxidation, DNA damage, membrane damage (Kamat et al., 2000; Kumar et al., 2001). A burst of free radical formation is demonstrated during cerebral ischaemia and reperfusion induced injury. Chlorophyll salt and the aqueous extract of *Baccopa monneria* and *Valeriana wallichii* exerts neuroprotective effects (Rehni et al., 2007).

2.2.2. Modifier of Genotoxic Effect

Hartman and Shankel (1990) reviewed inhibitors that directly interact with mutagen and carcinogen and sequester so that they may not have any harmful effect on body. These inhibitors act as interceptor molecules against mutagen and carcinogen. Interceptors are are proficient in binding to, or reacting with, mutagenic chemicals and free radicals, and serves as a first line of defense against mutagens and carcinogens (Hartman and Shankel, 1990). Following interception, the defense mechanism may either involve induction of detoxification enzyme or inhibition of carcinogen activating enzyme. Data on activity profiles of antimutagens has been reviewed *in vitro* and *in vivo* data by Waters et al. (1996). Among the various inhibitors reviewed, chlorophyllin (CHL) was identified as almost uniformly protective against a broad range of direct- and indirect-acting mutagens, including aflatoxins, polycyclic aromatic hydrocarbons, heterocyclic amines, alkylating agents and several miscellaneous compounds (Arimoto et al., 1993; Breinholt et al., 1995; Tachino et al., 1994; Negishi et al., 1997; Dashwood et al., 1992, 1996, 1998, Dashwood, 2000).

Although chlorophyll and its compounds has potential to act antimutagens *in vitro* (Negishi et al., 1989, Dashwood et al., 1995) however they have shown chemopreventive properties *in vivo* such as chemoprevention of aflatoxin B_1 (AFB_1) hepatocellular carcinoma (HCC) in rainbow trout model (Breinholt et al., 1995, Dashwood et al., 1998; Reddy et al., 1999 Pratt et al., 2006; Simonich et al., 2008; Castro et al., 2009) and in rodent model (Guo et al., 1995; Hasegawa et al., 1995, Simonich et al., 2007) and human intervention (Yu, 1995; Egner et al., 2001). Chlorophyllin has strong binding capacity to acridine, more effectively than resveratrol and xanthenes (Osowski et al., 2010), which prevents DNA-mutagen intercalation.

2.2.3. Inhibition of Cytochrome P450 Enzymes

Cytochrome P450 enzymes are involved in the removal of carcinogenic compounds from the body. However, in some cases they can also activate compounds consumed in food, converting procarcinogens to carcinogens. Aflatoxin B1 is not carcinogenic until converted to the electrophilic 8,9 – epoxide, which can form adduct with DNA.

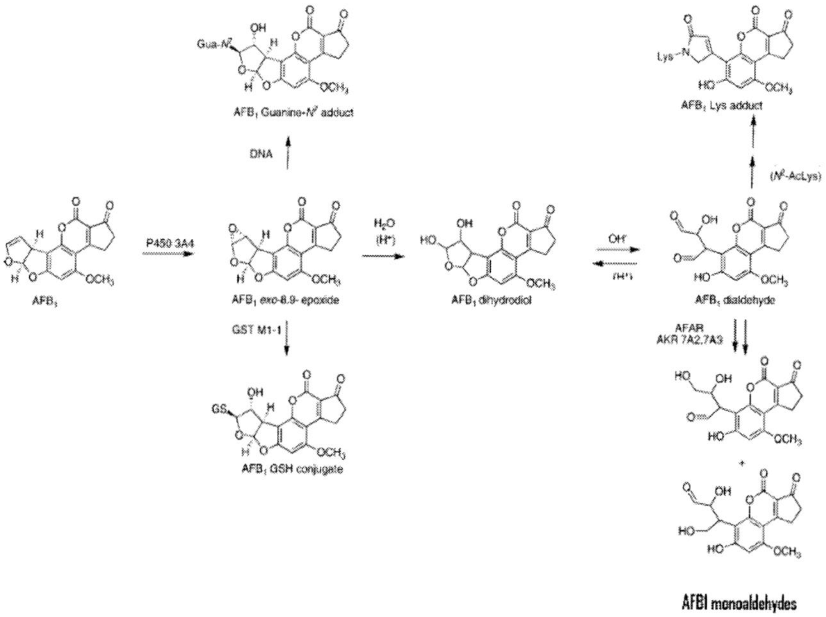

Figure 3. Metabolism of aflatoxin B1 in human (Guengerich et al., 2002, Guengerich, 2008).

The metabolic activation of AFB1 is mediated by cytochrome p450 (Tachino et al., 1994; Yun et al., 1995). Dietary supplementation of chlorophyllin has significantly reduced AFB-1 induced DNA damage in the liver of rainbow trout and rats (Breinholt et al., 1995). The major pathway in metabolism of aflatoxin B1 in human is presented in Figure 3 (Guengerich et al., 2002, Guengerich, 2008).

CYP1B1 is also implicated in tobacco smoke-related cancers in several organs. Tobacco smoke contains several procarcinogens, including polycyclic aromatic hydrocarbons (PAHs), nitrosamines and arylamines. PAHs can be activated into carcinogens by CYP1A1, CYP1A2 and CYP1B1. Benzo[a]pyrene (BP) is a potent pro-carcinogen and ubiquitous environmental pollutant. John et al. (2010) observed the induction and modulation of CYP1A1 and CYP1B1 and 10-(deoxyguanosin-N2-yl)-7,8,9-trihydroxy-7,8,9,10-tetrahydrobenzo[a]pyrene (BPdG) adduct formation in DNA from primary normal human mammary epithelial cell (NHMEC) strains. Maximum percent reductions of CYP1A1 and CYP1B1 gene expression and BPdG adduct formation were observed when cells were pre-dosed with chlorophyllin followed by administration of the carcinogen. Chlorophyllin is likely to be a good chemoprotective agent for a large proportion of the human population.

2.2.4. Induction of Phase II Enzymes

Induction of phase II response is recognized as an effective strategy for protecting cells against oxidants, electrophiles.Phase II enzyme include glutathione-S-transferase, UDP glucoronosyl transferase, sulfotransferase, and oxidoreductase. Phase II enzymes bind to oxygenated carcinogens making highly polar molecule that are excreted. Phase II enzymes decrease carcinogenicity by blocking carcinogen metabolic activation and enhancing carcinogen detoxification. Although the *Brassica* vegetables have long been known to contain potent inducers of mammalian phase 2 enzymes (Dinkova-Kostova et al., 2004), chlorophyllin may also increase the activity of the phase II enzyme, quinone reductase (Dingley et al., 2003). Chlorophylls, chlorophyllin and related tetrapyrroles are significant inducer of mammalian phase II cytoprotective genes, inducing the phase 2 enzyme NAD(P) H:quinone oxidoreductase 1 (NQO1) in murine hepatoma cells (Fahey et al., 2005). The drug metabolizing enzyme comprises phase I (oxidation, reduction and hydrolysis). Physiological balance between Phase I and Phase II enzymes, and their level of expression and genetic polymorphism might dictate the sensitivity or risk of individual exposed to carcinogenic species (Kensler, 1997).

2.2.5. Effect of Chlorophyll on Cell Differentiation, Cell Arrest and Apoptosis of Cancer Cells

Generally, growth rate of pre-neoplastic or neoplastic cells is fast than normal cell. Therefore, induction of apoptosis or cell cycle arrest can be an excellent approach to inhibit the promotion and progression of carcinogenesis. Distinct from apoptotic events in the normal physiological process, which are mainly mediated by interaction between death receptors and their relevant ligands (Jacks and Weinberg, 2002), many dietary supplements appear to induce apoptosis through the mitochondria-mediated pathways. The cytotoxic effects of chemotherapeutic compounds on neoplastic cells can be monitored by measuring their effect on mitochondria, caspases and other apoptosis – related proteins. Chlorophyllin induced apoptosis in HCT116 human colon cancer cells, via a cytochrome c–independent pathway (Diaz et al., 2003).

Progression through cell cycle is a sequential process that directs cells to pass through G1, S. G2 and M. There are G1-S/ or G2-M checkpoints that halts cell division whenever necessary. Cyclin dependent kinase (CDKs) CDK inhibitors governs the progression of the cell cycle. Cell cycle arrest induced by chemopreventive compounds potentially affects and blocks the continuous proliferation of tumorogenic cells. Lower doses of CHL also were observed to induce cell-cycle arrest and strongly altered markers of cell differentiation, such as E-cadherin (Carter et al., 2004).A recent study showed that human colon cancer cells undergo cell cycle arrest after treatment with chlorophyllin (Chimploy, 2009). The mechanism involved inhibition of ribonucleotide reductase activity. Ribonucleotide reductase plays a pivotal role in DNA synthesis and repair, and is a target of currently used cancer therapeutic agents, such as hydroxyurea (Chimploy et al., 2009).

3. Applications in Cancer Chemotherapy

Cancer development is a long term process that involves initiation, promotion and progression that ultimately leads to spread from one area of the body to another during the late metastasis stage. Current clinical therapies which include surgery, radiotherapy and chemotherapy are limited to particularly during metastasis phase. However, there is increasing body of evidences from epidemiological and pathological studies that certain dietary substances may prevent or slow down progression of cancer. Because advance metastasis stage cancer are almost impossible to cure, therefore, cancer chemoprevention and containment at early stage is highly desirable. Dietary chemopreventive agents seems to have variety of cellular and molecular mechanism that may inhibit carcinogenesis (blocking agent) or suppress promotion and progression of carcinogenesis (suppressive agent) or function as both. Many dietary substances such as retinoic acid, sulforaphane, curcumin, EGCG, apigenin, qurecetin, chrysin, silibinin, silymarin and resveratrol acts through induction of apoptosis. Potential mechanism underlying effectiveness of some of the dietery constituents is presented in Table 2. Many dietary compounds including chlorophyll possess cancer protective properties that include cellular detoxifying mechanism and antioxidant property that protects against cellular damage caused by environmental carcinogens or endogenously generated reactive oxygen species. These dietary substances can affect death signaling pathways which could prevent proliferation of tumor cells.

Table 2. Potential Mechanism of action of some of the dietery chemopreventive compounds (modified from Chen and Kong, 2005)

Class	Function	Compounds	Source
Cancer Blocking agent	Enhanced detoxification of chemicals	Indole-3-Carbinol	Cruciferous vegetables
		Chlorophyll and its derivatives	Green leafy vegetables
		Sulpforaphane	Cruciferous vegetables
		Curcumin	Turmeric
	Inhibit cytochrome P450	Isothiocyanates	Cruciferous vegetables
		Selenium	Nuts and meat
		Vitamin E	Vegetable oil
	Trap carcinogen	Flavonoids	Fruits and Vegetables
		Chlorophyllin	Commercial preparation from chlorophyll
Suppressive agents	Cell cycle disruption /or induce apoptosis	Chlorophyllin	Commercial preparation from chlorophyll
		EGCG	Green tea
		Quercetin	Onion and tomatoes
		Resveratrol	Grapes
		Curcumin	Turmeric
		Sulphoraphane	Cruciferous vegetables

Chlorophyll has a potential to act as chemopreventive agent. Clinical trials with chlorophyllin have reduced aflatoxin-DNA adducts in individuals at high risk for liver cancer (Kensler et al, 1998, Dingley et al., 2003). In another clinical trial on patient with fibroadenomastosis of breast cancer, the drug mamoclam- containing mega-3 polyunsaturated fatty acids, iodine and chlorophyll derivatives, produced from the brown sea alga laminaria, was effective in pain relief and breast cyst regression (Bezpalov et al., 2005).

Chlorophyll can assist with the effects of dietary and environmental exposure to carcinogens. Notable examples are the tobacco-related carcinogens (e.g., nitrosamines and polycyclic aromatic hydrocarbons-PAH), heterocyclic amines produced from sustained, high-temperature cooking of meats and the fungal food contaminants aflatoxins. Research indicates that chlorophyll reduces carcinogen binding to DNA in the target organ by inhibition of carcinogen activation enzyme or degradation of ultimate carcinogens with the target cells. *In vitro* and *in vivo* studies further substantiated medicinal cures offered by chlorophyll derivatives. Elucidation of the molecular mechanisms of chemical carcinogenesis provides insight into targets for chemoprevention. Microarray and proteonomics analysis have shown alteration at the level gene expression and protein. Recently research investigations showing involvement of transcriptional factors and their intervention by chlorophyll and its derivative seems to be an attractive approach showing precise action at benefits of chlorophyll at cellular and molecular levels.

4. Photosensitizer

Photodynamic therapy (PDT) is increasingly becoming accepted as a treatment option for a variety of cancer which is usually based on the photosensitisation of tumour cells with subsequent light exposure leading to death of the malignant cells. The most commonly used photosensitisers, such as the haematoporphyrin derivatives have a number of drawbacks - poor selectivity in terms of tumour drug accumulation and low extinction coefficients so that relatively large amounts of drug and/or light are needed in order to obtain a satisfactory phototherapeutic response. These have led to the development of a further generation of photosensitisers based on chlorophyll derivatives, which are characterized by increased phototoxicity and strong absorption which allows deeper light penetration into tissues, rapid tissue clearance and minimal extravasation from the circulation (Rosenbach-Belkin et.al., 1996).

Aminolevulinic acid (ALA), a building block of tetrapyrroles, synthesized during chlorophyll biosynthesis, has shown photodynamic destruction of cancer cells (Reibeiz et al., 2002). It is available as porphyric insecticides and show photodynamic property. It can be easily taken up by transformed cells, and is rapidly cleared from the circulatory stream within 48 hr of treatment (Reibeiz et al., 2002). Conjugating Cp 6 with histamine can help improve the effectiveness of PDT in oral cancer cells by enhancing its intracellular delivery (Parihar et al., 2010). "Radachlorin"(®), also known in the Bremachlorin, a composition of 3 chlorophyll a derivatives in an aqueous solution, was introduced into the Russian Pharmacopoeia. Iand may be commercialized as a prospective second-generation photosensitizer (Kochneva et al., 2010). Prostate-specific membrane antigen (PSMA), a validated biomarker for prostate cancer, has attracted considerable attention as a target for imaging and therapeutic applications for prostate cancer. PSMA inhibitor, i.e. conjugate of pyropheophorbide has been used for targeted PDT application and the mechanism of its mediated-cell death in prostate cancer: inducing apoptosis via activation of the caspase-8/-3 cascade pathway (Liu et al., 2010).

5. Contraindications and Safety

Natural chlorophylls are not known to be toxic, and no toxic effects have been attributed to chlorophyllin despite more than 50 years of clinical use in humans. Although few contraindications have been reported by some investigators ((Chernomorsky, 1988; Gogel et al., 1989; Kephart, 1995; Egner et al., 2003, Hendler et al., 2008)) but much serious ill effects are less known. There is lack of reports about the safety of chlorophyll or chlorophyllin supplements in pregnant or lactating women. Although there is no major contraindication reported so far as but in order to be used in modern, pharmacological aspects the way these medicine interacts with human system needs to be further explored so as to recommend its safe, effective and widespread use of in medicine.

6. Challenges of Potential Application Chlorophyll and Its Derivatives as Chemotherapeutic Agents

Dietary chemoprevetive compounds offer great potential in fight against cancer. The mechanistic insight into chemoprevention of carcinogenesis includes regulation of cell defensive and cell-death machineries. Though progress have been made in understating apoptosis, and cell cycle arrest in relation to chlorophyll and its derivatives, signaling pathways and gene expression events leading to pharmacological effects require further investigation. The ultimate goal is translation of the results of in vitro signaling and gene expression obtained in animal cell culture system /animal model to beneficial pharmacological effects which have several challenges that need to be overcome. One of the concern is induction of detoxifying enzymes by chemotherapeutic agents varies within human population. Since the cancer development is a long term process, there is need to explore suitable potency indicators to assess the effect of chemopreventive agents. Dietery chemopreventive agents may not possess pharmacologically active properties to be used as drug. Studies on pharmacokinetics and toxicity profile of chemopreventive agents are important in drug development. Synergistic effect chemopreventive agent in association with other efficacious drug molecule might enhance the efficacy. Ultimately to convert a dietery chemopreventive agents into a viable drug, a clear understanding in this area will provide impetus for future developments.

References

Agarwal A, Sharma RK, Nallella KP, Thomas AJ Jr, Alvarez JG and Sikka SC (2006) Reactive oxygen species as an indepen-dent marker of male factor infertility. *Fertil. Steril.* 86, 878–885.

Arimoto S, Fukuoko S, Itome C, Nakano H, Rai H, Hayatsu H (1993) Binding of polycyclic planar mutagens to chlorophyllin resulting in inhibition of the mutagenic activity. *Mutat. Res.* 287:293-305.

Bezpalov VG, Barash NIu, Ivanova OA, Semenov II, Aleksandrov VA, Semiglazov VF (2005). *Voprosy onkologii* 51::236-241.

Bowers WF (1947) Chlorophyll in wound healing and suppurative disease. *Am. J. Surg.* 73 :49- 55.

Breinholt V, Hendricks J, Pereira C, Arbogast D, Bailey G (1995) Dietary chlorophyllin is a potent inhibitor of aflatoxin B1 hepatocarcino-genesis in rainbow trout. *Cancer Res.* 55:57- 62.

Carter, O., George S. Bailey, and Roderick H. Dashwood (2004). The Dietary Phytochemical Chlorophyllin Alters E-Cadherin and β-Catenin Expression in Human Colon Cancer Cells. *J. Nutr.* 134: 3441S–3444S.

Castro, DJ, Christiane VL, Kay AF, Katrina MW, Bobbie-Jo, M., Robertson, W, Roderick H, Dashwood RH, George S, Bailey GS, Williams DE (2009) Identifying efficacious approaches to chemoprevention with chlorophyllin, purified chlorophylls and freeze-dried spinach in a mouse model of transplacental carcinogenesis. *Carcinogenesis* 30:315–320.

Chen, C, Kong, ANT (2005). Dietary cancer-chemopreventive compounds:from signaling and gene expression to pharmacological effect. *Trends in Pharmacological Sciences* 26 (6):318-326.

Chernomorsky SA, Segelman AB (1988) Biological activities of chlorophyll derivatives. *N. J. Med.* 85:669-673.

Chimploy, K., G. Dario Díaz, Qingjie Li, Orianna Carter,Wan-Mohaiza Dashwood,Christopher K. Mathews, David E. Williams,George S. Bailey, and Roderick H. Dashwood (2009). E2F4 and ribonucleotide reductase mediate S-phase arrest in colon cancer cells treated with chlorophyllin. *Int. J. Cancer* 125: 2086–2094.

Dashwood R, Yamane S, Larsen R (1996) Study of the forces of stabilizing complexes between chlorophylls and heterocyclic amine mutagens. *Environ. Mol. Mutagen.* 27:211-218.

Dashwood R (1997) Chlorophylls as anticarcinogens. *Int. J. Oncol.* 10:721–727.

Dashwood RH (2005) Modulation of heterocyclic amine-induced mutagenicity and carcinogenicity: An 'A-to-Z' guide to chemopreventive agents, promoters, and transgenic models. *Mutat. Res.* 511:89–112.

Dashwood RH, Guo D (1992) Inhibition of 2-amino-3-methylimidazo [4,5-f] quinoline (IQ)-DNA binding by chlorophyllin: studies of enzyme inhibition and molecular complex formation. *Carcinogenesis* 13:1121–1126.

Dashwood RH, Negishi T, Hayatsu H, Breinholt V, Hendricks JD, Bailey GS (1998) Chemopreventive properties of chlorophylls towards aflatoxin B1: a review of the antimutagenicity and anticarcinogenicity data in rainbow trout. *Mutat. Res.* 399:245–253.

Dashwood RH, Yamane S, Larsen R (1996) A study of the forces stabilizing complexes between chlorophylls and heterocyclic amine mutagens. *Environ. Mol. Mutagen.* 27:211–218.

Devasagayam, TPA , JC Tilak, KK Boloor, Ketaki S Sane, Saroj S Ghaskadbi, RD Lele (2004) Free Radicals and Antioxidants in Human Health:Current Status and Future Prospects. *Journal of Association of Physician of India* 52: 794-804.

Díaz GD, Li Q, Dashwood RH (2003) Caspase-8 and AIF mediate a cytochrome c-independent pathway of apoptosis in human colon cancer cells induced by the dietary phytochemical chlorophyllin. *Cancer Res.* 63:1254–1261.

Dingley KH, Ubick EA, Chiarappa-Zucca ML, Nowell S, Abel S, Ebeler SE, et al. (2003) Effect of dietary constituents with chemopreventive potential on adduct formation of a low dose of the heterocyclic amines PhIP and IQ and phase II hepatic enzymes. *Nutr. Cancer* 46: 212- 221.

Dinkova-Kostova AT, Fahey JW, Talalay P (2004) Chemical structures of inducers of nicotinamide quinone oxidoreductase 1 (NQO1). *Methods Enzymol.* 382:423--448.

Egner PA, Munoz A, Kensler TW (2003) Chemoprevention with chlorophyllin in individuals exposed to dietary aflatoxin. *Mutat. Res.* 523-524:209-216

Egner PA, Wang JB, Zhu YR, Zhang BC, Wu Y, Zhang QN, Quian GS, Kuang SY, Gange SJ, et al. (2001) Chlorophyllin intervention reduces aflatoxin-DNA adducts in individuals at high risk for liver cancer. *Proc. Natl. Acad. Sci.* 98:14601–14606.

Egner PA, Stansbury KH, Snyder EP, Rogers ME, Hintz PA, Kensler TW (2000) Identification and characterization of chlorin e(4) ethyl ester in sera of individuals participating in the chlorophyllin chemoprevention trial. *Chem. Res. Toxicol.* 13:900-906.

Elahi MM, Matata BM (2006) Free radicals in blood: evol-ving concepts in the mechanism of ischemic heart disease. *Arch. Biochem. Biophys.* 450:78–88.

Esten, MM. and Dannin, AG. (1950) "Chlorophyll therapy and its relation to pathogenic bacteria," Butler University Botanical Studies: Vol. 9, Article 21, pp 212-217 [http://digitalcommons.butler.edu/botanical/vol9/iss1/21].

Fahey JW, Stephenson KK, Dinkova-Kostova AT, Egner PA, Kensler TW, Talalay P (2005) Chlorophyll, chlorophyllin and related tetrapyrroles are significant inducers of mammalian phase 2 cytopro-tective genes. *Carcinogenesis* 26:1247- 1255.

Ferruzzi MG, Blakeslee J (2007) Digestion, absorption, and cancer preventative activity of dietary chlorophyll derivatives. *Nutrition Research* 27: 1–12.

Ferruzzi MG, Bfhm V, Courtney P, Schwartz SJ (2002) Antioxidant and antimutagenic activity of dietary chlorophyll derivatives determined by radical scavenging and bacterial reverse mutagenesis assays. *J. Food Sci.* 67:2589- 2595.

Gogel HK, Tandberg D, Strickland RG (1989) Substances that interfere with guaiac card tests: implications for gastric aspirate testing. *Am. J. Emerg. Med.* 7:474-480.

Guengerich, FP (2008). Cytochrome P450 and Chemical Toxicology. *Chem. Res. Toxicol.* 2008, 21, 70–83.

Guengerich, F. P., Arneson, K. O., Williams, K. M., Deng, Z., and Harris, T. M. (2002) Reaction of aflatoxin B oxidation products with lysine. *Chem. Res. Toxicol.* 15, 780–792.

Guo D, Horio D, Grove J, Dashwood RH (1995) Inhibition by chlorophyllin of 2-amino-3-methylimidazo [4,5-f] quinoline (IQ)-induced tumorigenesis in the male F344 rat. *Cancer Lett.* 95:161–165.

Hartman PE, Delbert MS (1990) Antimutagens and anticarcinogens: A survey of putative interceptor molecules. *Environmental and Molecular Mutagenesis.* 15:145–182.

Hasegawa R, Hirose M, Kato T, Hagiwara A, Boonyaphiphat P, Nagao M, Ito N, Shirai T (1995) Inhibitory effect of chlorophyllin on PhIP-induced mammary carcinogenesis in female F344 rats. *Carcinogenesis* 16:2243–2246.

Henderson BW, Bellnier DA, Greco WR, Sharma A, Pandry RK, Vaughan LA, et al. (1997) An in vivo quantitative structure-activity relationship for a congeneric series of pyropheophorbide derivatives as photosensitizers for photodynamic therapy. *Cancer Res.* 57: 4000- 4007.

Hendler SS, Rorvik DR (2008) PDR for Nutritional Supplements. 2nd ed. Montvale: Physicians' Desk Reference, Inc.

Jacks T, Weinberg RA (2002) Taking the study of cancer cell survival to a new dimension. *Cell* 111:923-925.

John K, Divi RL, Keshava C, Orozco CC, Schockley ME, Richardson DL, Poirier MC, Nath J, Weston A (2010). CYP1A1 and CYP1B1 gene expression and DNA adduct formation in normal human mammary epithelial cells exposed to benzo[a]pyrene in the absence or presence of chlorophyllin. *Cancer Lett.* 28:254-60.

Kamat JP, Boloor, KK, Devasagayam, TPA (2000).Chlorophyllin as an effective antioxidant against membrane damage in vitro and in vivo. *Biochim. Biophys. Acta* 1487:113-127.

Kensler TW (1997) Chemoprevention by inducers of carcinogen detoxification enzymes. *Environ. Health Perspect.* 105: 965-970.

Kensler TW, Groopman JD, Roebuck BD (1998) Use of aflatoxin adducts as intermediate endpoints to assess the efficacy of chemopreventive interventions in animals and man. *Mutat. Res.* 402:165-172.

Kephart JC (1955) Chlorophyll derivatives - their chemistry, commercial preparation and uses. *Econ. Bot.* 9: 3-38.

Knasmüller S, Nersesyan A, Mišík M, Gerner M, Mits W, Ehrlich V, Hoelzl C, Szakmary A (2008) Use of conventional and -omics based methods for health claims of dietary. *British Journal of Nutrition 99 E Suppl.*1.: ES2-ES52.

Kochneva EV, Filonenko EV, Vakulovskaya EG, Scherbakova EG, Seliverstov OV, Markichev NA, Reshetnickov AV (2010) Photosensitizer Radachlorin®: Skin cancer PDT phase II clinical trials. *Photodiagnosis Photodyn. Ther.* 7:258-267.

Kumar SS, Devasagayam TPA, Bhushan B, Verma NC (2001) Scavenging of reactive oxygen species by chlorophyllin: an ESR study. *Free Rad. Res.* 35:563-574.

Lanfer-Marquez UM, Barros RMC Sinnecker P (2005) Antioxidant activity of chlorophylls and their derivatives. *Food Research International* 38:885-891.

Larato DC, Pfau FR (1970) Effects of water soluble chlorophyllin ointment on gingival inflammation. *NY Dent. J.* 36:291- 293.

Lefer DJ, Granger DN (2000) Oxidative stress and cardiac disease. *Am. J. Med.* 109:315–323.

Li G, Slansky A, Dobhal MP, Goswami LN, Graham A, Chen Y, Kanter P, Alberico RA, Spernyak J, Morgan J, Mazurchuk R, Oseroff A, Grossman Z, Pandey RK (2005) Chlorophyll-a analogues conjugated with aminobenzyl-DTPA as potential bifunctional agents for magnetic resonance imaging and photodynamic therapy. *Bioconjug. Chem.* 16:32-42.

Liu T, Wu LY, Choi JK, Berkman CE (2010). Targeted photodynamic therapy for prostate cancer: inducing apoptosis via activation of the caspase-8/-3 cascade pathway. *Int. J. Oncol.* 36:777-784.

Negishi T, Arimoto S, Nishizaki C, Hayatsu, H (1989) Inhibitory effect of chlorophyll on the genotoxicity of 3-amino-1-methyl-5H-pyrido[4,3-b indole (Trp-P-2). *Carcinogenesis* 10:145-149.

Negishi T, Rai H, Hayatsu H (1997) Antigenotoxic activity of natural chlorophylls. *Mutat. Res.* 376: 97-100.

Osowski A, Pietrzak M, Wieczorek Z, Wieczorek J (2010) Natural compounds in the human diet and their ability to bind mutagens prevents DNA-mutagen intercalation. *J. Toxicol. Environ. Health A*. 73:1141-1149.

Parihar A, Dube A, Gupta PK (2010) Conjugation of chlorin p (6) to histamine enhances its cellular uptake and phototoxicity in oral cancer cells. Cancer Chemother Pharmacol 2010 [PMID:20978762].

Park YJ, Lee WY, Hahn BS, Han MJ, Yang WI, Kim BS (1989) Chlorophyll derivatives--a new photosensitizer for photodynamic therapy of cancer in mice. *Yonsei. Med. J.* 30:212-218.

Pasqualotto FF, Sharma RK, Kobayashi H, Nelson DR, Thomas AJ Jr, Agarwal A (2001) Oxidative stress in normo spermic men undergoing infertility evaluation. *J. Androl.* 22: 316–322.

Pratt MM, Reddy AP, Hendricks JD, Pereira C, Kensler TW Bailey GS (2006) The importance of carcinogen dose in chemoprevention studies: quantitative interrelationships between, dibenzo[*a,l*]pyrene dose, chlorophyllin dose, target organ DNA adduct biomarkers and final tumor outcome . *Carcinogenesis* 28: 611-624.

Reddy AP, Harttig U, Barth MC, Baird WM, Schimerlik M, Hendricks JD, Bailey GS (1999) Inhibition of dibenzo[a,l]pyrene-induced multi-organ carcinogenesis by dietary chlorophyllin in rainbow trout. *Carcinogenesis* 20: 1919-1926.

Rehni AK, Pantlya HS, Shri R, Singh M (2007) Effect of chlorophyll and aqueous extracts of Bacopa monniera and Valeriana wallichii on ischaemia and reperfusion induced cerebral injury in mice. *Indian Journal of Experimental Biology* 45:764-769.

Reibeiz, CA, Kolossov, VL, Briskin, D, Gawienowski, M(2002). Chloroplast biogenesis: chlorophyll biosynthesis heterogeneity, multiple biosynthetic routes and biological spin-offs. In: *Handbook of photochemistry and photobiology*, Vol. IV (ed. H.S. Nalwa: Foreword by Professor Jean-Marie Lehn, Nobel Laureate in chemistry), American Scientific Publishers, Los Angeles, USA, pp. 183-268.

Rosenbach-Belkin V, Chen L, Fiedor L, Tregub I, Pavlotsky F, Brunfeld V, Salomon Y, Scherz A. (1996). Serine conjugates of chlorophyll and bacteriochlorophyll: photocytotoxicity *in vitro* and tissue distribution in mice bearing melanoma tumors. *Photochem. Photobiol.* 64:174–181.

Sato M, Imai K, Kimura R, Murata T (1984) Effect of sodium copper chlorophyllin on lipid peroxidation. VI. Effect of its administration on mitochondrial and microsomal lipid peroxidation in rat liver. *Chem. Pharm. Bull.* 32:716-722.

Sato M, Imai K, Kimura R, Murata T (1985) Effect of sodium copper chlorophyllin on lipid peroxidation. VIII. Its effect on carbon tetrachloride-induced liver injury in rats. *Chem. Pharm. Bull.* 33: 3530-3533.

Sato M, Konagai K, Kimura R, Murata T (1983) Effect of sodium copper chlorophyllin on lipid peroxidation. V. Effect on peroxidative damage of rat liver lysosomes. *Chem. Pharm. Bull.* 31: 3665-3670.

Schumacker PT (2006) Reactive oxygen species in cancer cells: live by the sword, die by the sword. *Cancer Cell* 10:175-176.

Simonich MT, Egner PA, Roebuck BD, Orner GA, Jubert C, Pereira C, Groopman JD, Kensler TW, Dashwood RH Williams DE George S, Bailey GS (2007) Natural chlorophyll inhibits aflatoxin B1-induced multi-organ carcinogenesis in the rat. *Carcinogenesis.* 28:1294-1302.

Simonich, MT, McQuistan T, Jubert C, Pereira C, Hendricks JD, Schimerlik M, Zhu B, Dashwood RH, Williams DE, Bailey GS (2008) Low-dose dietary chlorophyll inhibits multi-organ carcinogenesis in the rainbow trout. *Food Chem. Toxicol.* 46:1014–1024.

Tachino N, Guo D, Dashwood WM, Yamane S, Larsen R, Dashwood R (1994) Mechanism of the in vitro antimutagenic activity of chlorophyllin against benzo[a]pyrene: studies of enzyme inhibition, molecular complex formation and degradation of the ultimate carcinogen. *Mutat. Res.* 308:191-203.

Von Kobbe C, May A, Grandori C, Bohr VA (2004) Werner syndrome cells escape hydrogen peroxide-induced cell pro-liferation arrest. *Federation of American Societies for Exp. Biol. J.* 18:1970–1972.

Waters MD, Stack HF, Jackson MA, Brockman HE, De Flora S (1996) Activity profiles of antimutagens: in vitro and in vivo data. *Mutat. Res.* 350:109–129.

Yoshikawa T, Toyokuni S, Yamamoto Y and Naito Y, (eds). *Free Radicals in Chemistry Biology and Medicine,* OICA International, London, 2000.

Young RW, Beregi JS (1980) Use of chlorophyllin in the treatment of geriatric patients. *J. Am. Geriatr. Soc.* 28:46-7.

Yu SZ (1995) Primary prevention of hepatocellular carcinoma. J. Gastroenterol. *Hepatol.* 10: 674–682.

Yun CH, Jeong HG, Jhoun JW, Guengerich FP (1995) Non-specific inhibition of cytochrome P450 activities by chlorophyllin in human and rat liver microsomes. *Carcinogenesis* 16:1437- 1440.

Index

A

abstraction, 10
acetone, 5, 18, 20, 23, 24, 28, 42, 156
acid, 4, 12, 13, 19, 33, 46, 48, 49, 91, 120, 178, 186, 188
acidity, 181
acne, 13, 36
activation energy, 10
active oxygen, 102
active site, ix, 44, 70
adaptation, 104, 107, 113, 126
advancement, x, 82, 91
aflatoxin, 7, 40, 183, 184, 187, 190, 191, 193, 195
age, 66, 142
aggregation, 7, 11, 14, 19, 35, 51, 74
agriculture, 113
air pollutants, 138, 174
air temperature, 164
alfalfa, 105
algae, vii, x, 3, 27, 48, 105, 116, 118, 120, 138, 171
aluminium, 108
amplitude, 53, 123, 154
anchoring, 46
angiosperm, 72
anti-cancer, vii, 1
antigen, 178, 188
antioxidant, 143, 181, 182, 186, 193
apoptosis, 181, 185, 186, 188, 189, 191, 193
apples, 134, 149
arginine, 101
aromatic hydrocarbons, 182, 184, 187
arrest, 181, 185, 189, 190, 195
arthritis, 13, 182
aspirate, 192
assessment, 96, 110, 138, 148, 153, 157, 171, 174
assessment procedures, 96
assimilation, 83, 90, 94, 95, 101, 106, 113, 122, 140, 142, 146
atmosphere, 139
atoms, ix, 4, 18, 43
attachment, ix, 43, 60, 63
autooxidation, 19

B

bacteria, 3, 20, 27, 68, 120, 121, 191
ban, 11
bandwidth, 53
barriers, 58
basal cell carcinoma, 13
base, 136, 139, 149
benefits, 178, 187
beverages, 178
bilirubin, 14
biological activity, 12
biological systems, 33, 45

biomarkers, 170, 194
biomass, 97, 98, 138
biomedical applications, vii, 30
biosphere, 16, 44, 139
biosynthesis, vii, ix, 43, 44, 45, 49, 51, 61, 62, 64, 65, 66, 67, 68, 70, 72, 75, 76, 77, 79, 84, 86, 91, 99, 188, 194
biosynthetic pathways, 65
biotic, 82, 92, 159, 170
bleaching, 19, 20, 22, 23, 24, 27, 28, 42, 59
blood, 12, 179, 181, 191
bonding, 5, 6, 48
bonds, 4, 5, 19
boreal forest, 140, 147
branching, 65
breakdown, 12, 18, 31, 35, 84, 99, 105, 107, 109, 174
breast cancer, 187
breeding, vii, x, 82, 83, 87, 88, 89, 96, 99, 101, 110, 111, 138, 146
building blocks, 84

C

cancer, vii, xi, 1, 7, 10, 11, 13, 31, 37, 177, 178, 181, 185, 187, 188, 189, 190, 191, 192, 193, 194, 195
carbohydrates, 11, 114
carbon, 83, 90, 95, 106, 123, 139, 146, 155, 195
carboxyl, 48
carcinoma, 12, 13, 36, 183, 196
carotene, 33
carotenoids, x, 21, 25, 26, 29, 63, 116, 122, 138, 148, 151, 156, 157, 162, 174
caspases, 185
catabolism, 84, 85, 99, 102, 103, 112, 113
catalysis, 70
catalytic activity, 77
chain propagation, 181
challenges, ix, xii, 81, 83, 99, 178, 189
charge coupled device, 96
chemoprevention, 183, 186, 187, 189, 190, 191, 194
chemopreventive agents, 178, 186, 189, 190

chemotherapeutic agent, xii, 178, 189
chemotherapy, 185
chlorine, 23
chloroplast, 33, 36, 71, 131
chromatography, 57, 63
chromatography analysis, 57
chromosome, 106
chronic diseases, 178
circulation, 181, 188
classes, 90, 145
classification, 10
cleavage, 84
climate, ix, 81, 82, 99, 108, 139, 164, 170, 172
climatic factors, 173
clinical trials, 193
cloning, 69, 113
closure, 95
coding, 68, 69, 84
coefficient of variation, 163, 165, 167
coenzyme, 69
colon, 185, 190, 191
color, 83, 98, 105, 116, 154
commercial, 179, 181, 193
communities, 31
community, 98, 99, 138, 141, 149
comparative analysis, 56
competition, 10, 118, 122
compilation, 38
complexity, 18, 64
composition, viii, 2, 16, 40, 64, 83, 138, 145, 147, 149, 188
compounds, 4, 10, 19, 29, 38, 84, 120, 179, 182, 183, 185, 186, 189, 190, 194
conception, xi, 152, 162
condensation, 91
conductance, 95
configuration, 24, 59, 71
congress, 144, 146
consensus, 141
constant rate, 124
constituents, 58, 146, 186, 191
construction, 44
consumption, 178
controlled exposure, 14

Index 199

controversial, 14
controversies, 30, 133, 136, 141
convention, 174
cooking, 187
cooperation, 13
coordination, 35, 48
copper, 100, 179, 195
correlation, 88, 95, 104, 130, 140, 141, 159, 168
cost, 89
coumarins, 120
crown, 111
crystalline, 51
cultivars, 95, 101, 108
culture, 189
curcumin, 186
cure, 178, 185
cures, 187
cycles, 154
cycling, 19
cyst, 187
cytochrome, 181, 184, 185, 186, 191, 196
cytochrome p450, 184
cytokinins, 91
cytoplasm, 46

D

damages, 17
decomposition, 19, 35, 37, 49, 104, 171
deconvolution, 55
decoupling, 139
deficiencies, 12, 82, 85, 86
deficiency, ix, 82, 94, 103, 113, 120, 144
degradation, xi, 18, 19, 20, 25, 28, 34, 36, 82, 83, 85, 91, 100, 105, 113, 152, 160, 168, 172, 174, 187, 195
dehydration, 105
density functional theory, 58
deposition, 172
deposits, 155, 157, 162, 168, 174
depression, 110
depth, xi, 136, 152, 154, 159, 168
derivatives, vii, xii, 1, 2, 3, 5, 10, 11, 12, 15, 31, 33, 36, 38, 39, 40, 144, 178, 180, 181, 182, 186, 187, 188, 189, 190, 192, 193, 194
dermatitis, 181
dermatology, 12
destruction, 26, 157, 160, 165, 168, 169, 173, 174, 188
destruction processes, 168, 174
detectable, 19, 20
detection, 34, 35, 54, 139, 141, 147, 148, 149
detergents, 40
determinism, 106
detoxification, 182, 184, 186, 193
deviation, 88
diatoms, 118
diet, 7, 194
differential spectroscopy, 55
diffusion, 39
digestion, 181
digital cameras, 101
diodes, 13
discrimination, 139, 145
diseases, 13, 90, 178, 181
dissolved oxygen, 154
distribution, 114, 128, 131, 134, 136, 137, 143, 153, 154, 158, 159, 162, 163, 164, 167, 168, 169, 170, 174, 175, 195
divergence, 63
diversity, 86, 91
donors, 54, 118
double bonds, 5, 19
drought, ix, 82, 85, 90, 92, 95, 98, 100, 101, 102, 103, 104, 105, 107, 108, 110, 111, 114, 138, 148
drug delivery, 12, 13
drugs, xii, 13, 39, 177, 178
dry matter, xi, 152, 155, 157, 159, 160, 161, 162, 165, 167, 168, 169
drying, 157
dyes, 7, 131, 146
dynamism, 159

E

E-cadherin, 185

ecosystem, 163, 164, 169, 170, 172
elaboration, viii, 2
electron, 5, 10, 18, 20, 23, 35, 37, 41, 53, 54, 57, 60, 66, 94, 95, 108, 110, 118, 120, 122, 126, 145, 181
electrons, 18, 60, 94, 120, 122, 124
elucidation, 103
e-mail, 115
emission, vii, viii, x, 2, 9, 11, 21, 23, 27, 50, 72, 115, 116, 118, 119, 120, 122, 125, 127, 128, 129, 131, 132, 133, 135, 136, 138, 139, 141, 143, 145, 146, 147, 148
emulsions, viii, 2, 29
encoding, 90, 91
energy, vii, 1, 9, 10, 16, 20, 39, 49, 50, 62, 63, 83, 92, 116, 117, 118, 120, 121, 122, 124, 140
energy transfer, 10, 49, 50, 62, 63, 117, 118, 122
enzyme, ix, 44, 45, 46, 48, 49, 51, 57, 59, 70, 75, 76, 77, 91, 113, 182, 184, 187, 190, 195
enzymes, 15, 29, 72, 84, 90, 123, 181, 183, 184, 189, 191, 193
epithelial cells, 192
equilibrium, 53, 56
equipment, 92
ester, 18, 180, 191
ethylene, 91, 105
eukaryotic, 122
eukaryotic cell, 122
evidence, 53, 54, 58, 73, 136, 171
evolution, 70, 93, 129, 138
excitation, vii, 1, 8, 19, 50, 58, 63, 64, 116, 125, 128, 131, 132, 133, 135, 138, 142, 145, 147, 148
exciton, 8, 9, 15
exclusion, 103
experimental condition, 28
exposure, 14, 15, 16, 29, 36, 40, 187
extinction, 187
extraction, 156
extracts, 32, 63, 64, 65, 116, 156, 194
extravasation, 188

F

fatty acids, 187
fiber, 13
fiber optics, 13
field crops, 113
financial, 141
financial support, 141
Finland, 140
fish, 171
flexibility, 59
flights, 141
flooding, 91, 114
fluctuations, 154, 164, 172
fluorimeter, 144
food, ix, 81, 82, 90, 91, 103, 178, 181, 183, 187
food industry, 90
food production, 82
force, 16
formation, viii, ix, xi, 3, 11, 19, 20, 23, 24, 27, 28, 44, 45, 46, 51, 52, 53, 54, 55, 56, 57, 58, 59, 60, 62, 63, 64, 65, 66, 67, 70, 73, 76, 77, 78, 79, 152, 164, 171, 181, 182, 184, 190, 191, 192, 195
free energy, 121
free radicals, viii, 2, 3, 17, 58, 77, 181, 182
freezing, 59, 104
fruits, x, 104, 116, 138, 143, 178
fucoxanthin, 118, 138
functional food, 178
fungi, 85
fusion, 69, 105

G

genomics, vii, x, 82, 83, 89, 99, 100, 106
germination, 107
gingival, 193
glutathione, 113, 184
glycerol, 19
glycine, 84
graph, 52
green alga, 3, 48, 118, 171

greening, 49, 61, 65, 68, 73, 74, 78, 79, 145
growth, 14, 40, 73, 82, 90, 98, 99, 100, 103, 105, 111, 113, 185
guidelines, 35

H

harvesting, ix, 3, 17, 33, 38, 39, 41, 44, 61, 62, 63, 66, 74, 83, 98, 121, 122, 138
healing, viii, 2, 14, 16, 29, 178, 190
health, 7, 102, 137, 178, 179, 193
heart disease, 191
heavy metals, 138
heme, 4, 13, 179
hemoglobin, 4, 179
hepatocellular carcinoma, 12, 36, 183, 196
hepatoma, 184
herbicide, 100
heritability, 88, 89
heterogeneity, xi, 96, 98, 152, 162, 167, 175, 194
hexane, 20, 22, 23, 24, 28, 42
histamine, 188, 194
histogram, 162
history, 33, 178
hormones, 11
human, 7, 12, 32, 36, 68, 181, 183, 184, 185, 188, 189, 191, 192, 194, 196
hybrid, 104
hydrogen, ix, 10, 43, 45, 47, 50, 60, 62, 66, 181, 195
hydrogen abstraction, 10
hydrogen atoms, ix, 43
hydrogen peroxide, 181, 195
hydrolysis, 184
hydrophilicity, 11
hydroquinone, 41
hydroxyl, 34, 181
hypoxia, 170

I

ideal, 90, 131
identification, 35, 36, 40, 45, 68, 98, 106

idiopathic, 182
illumination, 49, 50, 51, 52, 53, 54, 57, 60, 61, 63, 64, 65, 75, 78
image, 96, 98, 141
image analysis, 98
imagery, 110
images, 98, 108, 111, 118
immune function, 182
in vitro, 5, 12, 16, 18, 28, 30, 34, 35, 37, 42, 60, 67, 69, 73, 76, 101, 182, 183, 189, 193, 195
in vivo, 5, 16, 25, 30, 32, 34, 37, 45, 50, 51, 54, 55, 56, 59, 67, 73, 74, 112, 128, 129, 136, 142, 143, 149, 182, 183, 187, 192, 193, 195
index numbers, 3
India, 81, 177, 191
indirect effect, 105
individuals, 187, 191
inducer, 184
induction, 95, 101, 113, 129, 181, 182, 184, 185, 186, 189
industry, 90, 101
infertility, 182, 189, 194
inflammation, 193
ingredients, 29, 178
inheritance, 91, 103
inhibition, 41, 48, 95, 128, 181, 182, 185, 187, 189, 190, 195, 196
inhibitor, 188, 190
initial state, 127
initiation, 90, 91, 185
injury, iv, 112, 182, 194, 195
integration, 63
integrity, 41
internalization, 12, 31
intervention, 91, 183, 187, 191
iodine, 187
iron, 109, 179
irradiation, vii, 17, 19, 21, 22, 23, 24, 26, 27, 28, 29, 38, 40, 42, 52, 126, 128, 129, 130, 135, 137, 140
isolation, 32
issues, 12

K

ketones, 42
kill, 182
kinetics, 20, 24, 56, 75, 104, 123, 124, 126, 127, 128, 129, 130, 137

L

lactic acid, 19
lakes, 157, 158, 160, 170, 171, 172, 173
lasers, 13
lead, 14, 15, 24, 29, 61, 112, 169
leakage, 181
lesions, viii, 2, 14, 15, 29, 85
leukemia, 13
life cycle, 18
lifetime, 8, 36, 74, 144
ligand, viii, 2, 5, 14
liquid chromatography, 57
liver, 184, 187, 191, 195, 196
localization, 37
loci, x, 82, 85, 100, 101, 102, 107, 109, 113, 114
locus, 100, 106, 112
low temperatures, 49, 50, 51, 52, 53, 54, 56, 57, 60, 94, 128, 137
luminescence, 146, 149
lysine, 46, 192

M

machinery, 44, 90
macromolecules, 12, 31
magnesium, 18, 84, 179
magnetic resonance, 40, 193
magnitude, 64, 130, 131, 136
majority, 84, 85, 167
malignant cells, 187
maltose, 69
man, 153, 154, 193
management, 109, 112
manganese, 34
manipulation, 91
mapping, 100, 102, 103, 109, 174
marine environment, 32, 41
mass, 46, 157, 159, 160
materials, 14, 126, 131
matrix, 5, 11, 17, 25, 29, 181
matrixes, 120
matter, xi, 66, 136, 152, 153, 155, 157, 159, 160, 161, 162, 163, 164, 165, 166, 167, 168, 169, 171, 173, 175
meat, 186
medical, 12
medicine, xi, 7, 10, 177, 178, 180, 188
melanin, 14
melanoma, 195
membranes, 12, 17, 30, 33, 36, 46, 47, 67, 78, 83
mesophyll, 90, 97
metabolism, 18, 36, 83, 84, 85, 86, 95, 106, 123, 181, 184
metabolites, 84
metabolizing, 184
metals, 138
metastasis, 185
meter, 92, 107, 108, 109, 112
methanol, 37
methodology, 140
mice, 194, 195
microsomes, 196
mission, 144
missions, 142
mitochondria, 185
mixing, 154
modern science, xii, 178
modifications, 11, 17, 48
monomers, 35
morphology, 114
mosaic, x, 151, 152, 153, 154, 159, 167, 169, 170
motif, 113
mutagen, 179, 182, 183, 194
mutagenesis, 68, 192
mutant, 47, 48, 70, 71, 72, 73, 78, 113, 131
mutation, 85, 102
mutations, 85
myoglobin, 4

N

naming, 21
nanoparticles, 13
necrosis, 7
neglect, 29
nicotinamide, 47, 59, 191
nitrogen, 18, 75, 96, 102, 103, 106, 108, 109, 110, 114, 144, 155, 174
nitrosamines, 184, 187
nitroxide, 34
normal development, 101
nucleic acid, 181
nucleus, 46
nutrient, ix, 82, 86, 92, 94, 120, 138, 155
nutrient concentrations, 155
nutrients, 95, 169

O

oil, viii, 2, 3, 29, 186
oligomers, 20
optical properties, 101
organ, 40, 187, 194, 195
organelles, ix, 3, 5, 12, 13, 25
organic compounds, 120
organic matter, 153, 157, 159, 160, 165, 168, 169, 171, 173
organic solvents, 11
organism, x, 91, 116, 122, 123, 127
organs, 12, 30, 184
ornamental plants, 90
ox, 184
oxygen, 7, 10, 17, 19, 25, 26, 32, 33, 34, 35, 36, 38, 86, 102, 103, 107, 120, 121, 139, 140, 154, 168, 173, 178, 181, 186, 189, 193, 195
oxygen absorption, 140
ozone, viii, 2, 16, 106, 143

P

pain, 14, 29, 187
parallel, ix, 44, 53, 59, 61, 65, 66, 67, 89
pathways, ix, 9, 10, 15, 25, 30, 33, 44, 55, 59, 63, 65, 67, 84, 100, 118, 185, 186, 189
peat, 155, 159, 162, 166, 168
peptidase, 46
peptide, 46
peptides, 11
permeable membrane, 14
permission, iv, 50, 132, 133, 134, 135
permit, 98
peroxidation, 30, 36, 181, 182, 195
peroxide, 19, 181, 195
pharmaceutical, vii, 2, 39
pharmaceuticals, 14, 32
pharmacokinetics, 189
phenol, 47
phenotype, 96, 112
phenotypes, 88
phosphate, 91, 102
phosphorescence, 9, 35, 41
phosphorus, 108, 114, 155, 174
photobleaching, 24, 31, 33, 37
photodegradation, 19, 32, 36, 38
photons, 6, 9, 20, 25, 26, 122
photooxidation, 17, 18, 19, 21, 31, 34, 37, 40
photosensitivity, 13
physics, 39, 41
physiology, x, 82, 83, 96, 110, 116
phytoplankton, xi, 25, 30, 31, 35, 138, 145, 146, 149, 152, 153, 155, 162, 164, 165, 167, 168, 169, 171, 172, 173
plankton, 155, 163, 165, 168
plasma membrane, 12, 13
plastid, 90
platform, 100
playing, viii, 2
polar, 18, 20, 23, 74, 184
polarization, 50
pollutants, 18, 138, 174
pollution, 82, 86, 174
polycyclic aromatic hydrocarbon, 182, 184, 187
polymers, 11
polymorphism, 184

polypeptides, 68
polyunsaturated fat, 187
polyunsaturated fatty acids, 187
pools, 49
population, 7, 37, 88, 107, 108, 184, 189
porphyrins, 6, 9, 12, 13, 14, 15, 116
positive correlation, 168
potato, 95, 104, 145
preparation, iv, 15, 39, 186, 193
preservation, 168, 169, 171, 172, 174
prevention, 15, 196
principles, 12, 41
probability, 118, 162
probe, 27, 59, 106
proliferation, 182, 185, 186
promoter, 91, 108
propagation, 181
prostate cancer, 13, 188, 193
protection, 26, 29, 32
protein oxidation, 182
proteins, 4, 11, 15, 36, 39, 46, 49, 75, 78, 90, 118, 181, 185
protons, 60
psoriasis, 13

Q

quanta, 93
quantification, 89
quantum yields, 10, 52, 146
quartz, 132
quercetin, 120
quinone, 41, 94, 122, 184, 191

R

radiation, 11, 16, 17, 19, 24, 30, 31, 33, 34, 35, 36, 37, 40, 41, 93, 96, 106, 119, 124, 155
radiotherapy, 185
rash, 13, 29
reading, 112
receptors, 185
recession, 164
recommendations, iv
recovery, 109, 147
recovery process, 109
reflectivity, 140
regeneration, 78
regression, 187
relaxation, 56
relevance, 7, 11, 82, 111, 148
relief, 14, 153, 154, 187
remote sensing, 139, 140, 141, 144, 148, 149
repair, 14, 31, 34, 185
reparation, 39, 193
requirements, 47, 96
researchers, 52, 56, 57, 61
residues, 12, 46, 48, 68, 70
resistance, 26, 90, 101, 102, 103, 107, 108, 110, 114
resolution, 98, 106, 148
respiration, 31, 33, 35, 37, 38, 39, 41
response, viii, x, 2, 16, 17, 18, 23, 27, 29, 33, 37, 82, 88, 89, 90, 92, 93, 95, 114, 127, 134, 135, 149, 170, 184, 187
restenosis, 13
restoration, 14, 62
resveratrol, 183, 186
retrovirus, 13
rheumatoid arthritis, 182
ribonucleotide reductase, 185, 190
ribosomal RNA, 90
rights, iv
rings, 4, 6, 18, 51
risk, 184, 187, 191
room temperature, x, 53, 55, 56, 58, 59, 60, 65, 76, 115, 120, 125, 127, 128, 129, 130, 136, 141, 144
routes, 61, 64, 66, 71, 194
rowing, 79
rules, 4, 8
runoff, 154

S

safety, 188

Index

salinity, 82, 86, 92, 95, 103, 106, 108, 109, 111, 112
salt tolerance, 107, 114
salts, 179
saturation, 154
scattering, 16, 97, 131, 144
science, x, xii, 35, 36, 82, 99, 178, 179
seasonal changes, 164, 167
seasonal flu, 164
second generation, 10
seed, 51, 74
seedlings, 71, 78, 100, 105
selectivity, 13, 187
senescence, 18, 36, 84, 85, 89, 90, 91, 98, 99, 101, 102, 103, 104, 105, 107, 108, 112, 113, 114, 138
sensing, 139, 140, 141, 142, 144, 147, 148, 149
sensitivity, 7, 10, 13, 14, 24, 30, 39, 92, 96, 101, 104, 143, 184
sensitization, 10, 15
sepsis, 181
sequencing, 69, 89
shade, 136
shock, 65, 79
shoots, 100, 106
showing, 93, 187
side chain, 32, 48, 83, 84
side effects, 14
signaling pathway, 186, 189
signals, 41, 96, 139
signs, 7, 32
silica, 14
silkworm, 36
simulation, 150
skeleton, 5
skin, viii, 2, 7, 13, 14, 15, 16, 29, 32, 41, 150
society, 99
sodium, 179, 195
software, 98
solution, 18, 19, 21, 23, 25, 26, 27, 36, 37, 40, 116, 118, 141, 188
solvents, 5, 11, 19, 20, 24, 41, 74
spectrophotometric method, 153, 156

spectrophotometry, 76
spectroscopy, 11, 27, 34, 38, 39, 40, 47, 49, 52, 53, 55, 56, 58, 59, 60, 61, 64, 74, 76, 118, 146
spin, 9, 34, 194
spongy tissue, 136
stability, vii, 1, 3, 17, 24, 26, 28, 29, 36, 101, 181
standard deviation, 88
starvation, 96, 138
stoichiometry, 135
stomata, 111, 123
stomatitis, 12, 36
storage, 98, 138
stratification, x, 151, 153, 154, 168
structural changes, 62
structural gene, 102
structure, vii, x, 4, 5, 17, 25, 26, 32, 35, 41, 46, 47, 48, 49, 59, 62, 65, 66, 68, 69, 73, 78, 83, 103, 141, 151, 153, 172, 179, 181, 192
substitution, 5
substrate, 10, 46, 48, 67, 70, 71
succession, 5
sugar beet, 96, 105, 144
sulfur, 105, 109, 120
sulphur, 96
supervision, 153
supplementation, 184
surface layer, 157, 164, 168
survival, 71, 192
susceptibility, 101, 107, 148
suspensions, 24, 78, 136
symmetry, 5
symptoms, 85, 98
syndrome, 91, 195
synthesis, 13, 42, 44, 49, 51, 61, 64, 65, 66, 67, 69, 71, 74, 83, 84, 85, 90, 120, 179, 185

T

target, 7, 11, 12, 91, 141, 185, 187, 188, 194
techniques, x, 11, 17, 27, 39, 82, 89, 92, 99, 102, 109

technologies, 7, 89, 98, 112
technology, 12, 31, 89, 96, 98, 103, 141
terraces, 154
territory, 154
testing, 192
theoretical approaches, x, 116
therapy, xi, 7, 10, 12, 13, 14, 29, 31, 32, 35, 37, 38, 39, 40, 177, 178, 179, 187, 191, 192, 193, 194
thermodynamic parameters, 58
time periods, 28
tissue, viii, x, 2, 7, 11, 14, 99, 115, 130, 131, 181, 187, 195
tobacco, 105, 107, 108, 112, 114, 184, 187
tobacco smoke, 184
toxic effect, 188
toxicity, 10, 86, 94, 144, 189
traits, vii, x, 82, 83, 85, 86, 88, 89, 98, 99, 110
transformation, 48, 53, 55, 58, 70, 73, 75, 79, 114, 173, 174
transformations, ix, 25, 44, 53
translation, 49, 189
transparency, 154, 157, 165
transport, 37, 41, 94, 95, 113, 145, 181
treatment, 7, 12, 13, 14, 36, 185, 187, 188, 195
trial, 187, 191
trophic state, x, 151, 153, 157, 162, 163, 168, 169, 170, 174
tryptophan, 12
tumor, 11, 37, 38, 186, 194
tumor cells, 186
tumorigenesis, 192
tumors, 7, 12, 13, 41, 195
tunneling, 58, 77
tyrosine, ix, 44, 46, 67, 121

U

ultrastructure, 131
uniform, 18

V

validation, 131
valuation, 194
variations, 128, 138, 172
varieties, 86, 95, 104, 113
vehicles, 42
viral diseases, 13

W

wavelengths, x, 3, 6, 10, 25, 31, 55, 96, 116, 127, 128, 129, 132, 133, 135, 146, 157
wheat germ, 110
wood, 85
wound healing, 14, 178, 190

X

xanthophyll, 143, 144

Y

yield, 11, 51, 53, 54, 59, 63, 82, 88, 89, 98, 105, 109, 110, 113, 114, 122, 124, 125, 126, 129, 130, 144, 145

Z

zinc, 70, 113